SxI – Springer for Innovation /
SxI – Springer per l'Innovazione

Volume 14

More information about this series at http://www.springer.com/series/10062

Massimiliano Granieri
Andrea Basso

Editors

Capacity Building in Technology Transfer

The European Experience

 Springer

Editors
Massimiliano Granieri
Department of Mechanical and Industrial
    Engineering
University of Brescia
Brescia
Italy

Andrea Basso
MITO Technology
Milan
Italy

ISSN 2239-2688   ISSN 2239-2696   (electronic)
SxI – Springer for Innovation / SxI – Springer per l'Innovazione
ISBN 978-3-319-91460-2   ISBN 978-3-319-91461-9   (eBook)
https://doi.org/10.1007/978-3-319-91461-9

Library of Congress Control Number: 2018942161

Printed on acid-free paper

This Springer imprint is published by the registered company Springer International Publishing AG
part of Springer Nature
The registered company address is: Gewerbestrasse 11, 6330 Cham, Switzerland

Acknowledgements

This book has been made possible by the efforts of many people that were involved in the European project ProgressTT, as well as other friends and colleagues that are engaged on a daily basis with the hard work of technology transfer. To all of them goes our gratitude.

A few names need to be recalled here: Laura McDonald and her team at ASTP-Proton headquarters in Leiden, together with Marta Catarino and Christian Stein; the experts of ProgressTT that worked with a selected number of technology transfer offices across Europe, to implement the capacity building strategy, whose names are Dominic De Groote, Christophe Haunold, John McManus, Michel Berg, Michel Morant, Sue Sundstrom, Tom Flanagan, Alan Kennedy, Shulamit Hirsch, Tom Hockaday, Lisa Cowey, Christi Mitchel, Dirk Groenewegen. Many thanks to all of them.

The European Union is often criticized, and the number of contrarians seems to be increasing. But the European Union is also challenging some of our comfort zones, and it is showing what is the potential of working together within professional and human communities. This book builds on that unique experience and counts the story of a different Europe.

Last but not least, we are indebted to Maria Cristina Acocella and Francesca Bonadei at Springer, always open to projects in science, technology, and innovation, but above all supportive and patient with authors and editors.

Contents

Part III   New Methodologies for Capacity Building

Building Capacity Building in Technology Transfer. An Introduction

Massimiliano Granieri and Andrea Basso

Abstract

This chapter provides an overview of the book and the underlying project about building capacity for technology transfer in Europe. Notwithstanding the robust scholarly production on technology transfer and the role of technology transfer offices, very rarely scholars and practitioners dealt with the intersection between technology management and organization theory and the implications for growing the capacity of a technology transfer organization to improve its performance.

An extensive stream of literature in technology management and organization deals with the role and the performance of technology transfer offices (TTOs). TTOs at universities, research hospitals, research foundations and other public research organizations play a crucial role in supporting the processes that are required to bring research results to market, through cooperative agreements, licenses, sales and, in some cases, through the creation of spin-off companies.

Seduced by the performance of large TTOs in the United States, particularly as a (real or alleged) consequence of the Bay-Dole Act, a worldwide movement exists that seeks to understand if and how that performance can be reproduced elsewhere. Sure enough, the ability to create impact and generate significant financial returns is not something a TTO can do irrespective of the quality of the background research, the presence of critical masses, the availability of financial resources and an industrial environment ready to absorb the technology created. In this respect, the American example should be seen in its entirety rather than just selectively.

M. Granieri (✉)
DIMI, University of Brescia, Brescia, Italy
e-mail: massimiliano.granieri@unibs.it

A. Basso
MITO Technology, Milan, Italy
e-mail: andrea.basso@mitotech.eu

© Springer International Publishing AG, part of Springer Nature 2019   1
M. Granieri and A. Basso (eds.), *Capacity Building in Technology Transfer*,
SxI – Springer for Innovation / SxI – Springer per l'Innovazione 14,
https://doi.org/10.1007/978-3-319-91461-9_1

And nonetheless, in the United States as elsewhere, TTOs and the professionals involved, other things being equal, are the critical factor of success.

While managerial literature investigates the determinants of TTOs performance, there has been also a constant attention by policy makers at regional, domestic and European level to identify policy measures that can improve the activity of TTOs and generate a significant return on investment for R&D expenditures. For more than two decades now, governments and institutions experimented a combination of initiatives around what it is now called, at least in part, the Third Mission, although the expression can be misleading, since technology transfer represents a smaller portion of all the activities of social and economic engagement by universities and public research organizations.

Much of the initiatives and the policy measures launched revolved around the idea of professional training, under the assumption that the performance of TTOs is largely influenced by skills and competences, that professionals typically lack, and that academic courses typically do not offer. Professional training and, later on, professional certification, were seen as a way to gain recognition for the technology transfer profession and to improve the internal legitimacy of the TTO, that is ordinarily needed for an accomplished and respected manager to propose and drive any change in the organization that advances technology transfer activities. Such change also includes the setting of the internal goal to have more human resources to staff a good office and, first and foremost, setting the agenda for the research community that also embraces technology transfer and it is not limited to the traditional publishing and teaching duties.

Professional training only partially solved the problem and while its importance and impact cannot be underestimated, clearly additional efforts are required to raise the overall performance of TTOs in Europe.

One of the main observations at both practice and scientific level was that the process of transferring technology to the market needs, soon or later, complementary financial resources, to deal with the lower maturity level of technology (now often measured in terms of technology readiness level, TRL) of public funded research results, as well as to validate such results at prototype level (proof-of-concept or proof-of-principle) or to move further towards the market through the creation of a company. Higher levels of maturity in technology development trigger different kinds of early stage/seed investors; for lower level of maturity, the combination of technological risks and asymmetric information leaves a great deal of promising technologies unexploited or underexploited. This situation has been clear to the European Commission since the launch of Europe 2020, when one of the action items was to identify any measure that would stimulate the technology market in Europe and solve the problem of the so-called sleeping patents, most of which of academic origin.

To attract additional resources and launch development programs for early stage technologies, there is a common belief that TTOs should also have a role in managing the several steps needed to access such resources and to handle them, while supporting inventors and faculty in the proof-of-concept stage.

The perception that the majority of TTOs in Europe (with the exclusion of the usual suspects) would not be able to efficiently manage financial resources for pre-seed investments and the willingness to fix the market failure by securing such financial resources for universities and TTOs had induced the European Commission to envisage two back-to-back pilot measures within the Work Programme 2012–2014 of Horizon 2020.

The creation of a Technology Transfer Financial Facility with an initial endowment of about 64 million Euros was supposed to be launched with resources also from the European Investment Group and, in particular, through the European Investment Fund and its Technology Transfer division. Later, those resources ended up in the Horizon 2020 Access to Risk Finance Work Programme 2016–2017, but their intended use remained the same. As of the writing of this book, those resources are flowing to intermediaries through national initiatives with the purpose of fostering technology transfer activities along the public-private trajectory.

But even before providing financial resources, the Commission clearly meant to take a step back and make sure the candidate TTOs and other intermediaries for the financial resources would be up to their role. In this respect, the Work Programme introduced, for the first time in Horizon 2020 and, to a larger extent, in Europe, a capacity building pilot action for technology transfer (hereinafter, CBTT) with the aim to go beyond pure training and professional development and to steadily empower TTOs with the capacity to manage the technology transfer process in a more efficient way.

A consortium of eight public and private entities, named PROGRESS-TT, was awarded the resources of CBTT and started its 3-year journey to identify and support a number of TTOs with the purpose to increase their overall technology transfer capacity and make them able to deal with financial investors and providers. The change in perspective with PROGRESS-TT was dramatic. For the first time, the goal of increasing the TTOs performance was not pursued by acting at individual level (the professional) but at organizational level, putting the TTO as such at center stage and crafting specific actions for building and strengthening the office capacity in several dimensions. Moreover, the idea of leveraging on training programs was expanded and somehow replaced by those of coaching and mentoring both individual TTOs and clusters of TTOs willing to join forces and reach critical masses within given regional contexts.

While this book is not part of PROGRESS-TT foreground and, as such, not one of the project's outcome, it draws partly from its intermediate and final results and tries to build a conceptual framework for capacity building in technology transfer that could be applied in the future for other actions. It is not by chance that when PROGRESS-TT was already delivering its activities, another capacity building actions was launched by the European Commission (DG Regio) to foster the TTOs performance in the Western Balkans.

The structure of the book and its chapters reflect the purpose of witnessing the creativity of PROGRESS-TT in outlining and accomplishing an otherwise poorly defined notion of capacity building. Moreover, it aims at putting capacity building in the context both of all the policy actions that so far have been implemented to improve the performance of TTOs and in the current scholarly debate on technology transfer and on the use of financial resources to accelerate the translation of research results into new products and services.

In the first chapter, Patrick McCutcheon endeavors to sketch the full line of policy interventions that have been launched over the years to fix the market failures of technology transfer, strengthen the background conditions, and endow TTOs with all the necessary resources to better perform at European level and in the international setting. The chapter provides and impressive account of the intensity and motivation of the European Commission in supporting the world of technology transfer while pursuing an R&D policy that is progressively interested in the impact of the results and not just on theoretical, yet crucial, outcomes.

Lutz Maicher, Katja Dralle Mjos and Liina Tonisson provide an initial overview of the determinants of TTOs performance and success with the purpose of highlighting the space and the directions that in principle exists for capacity building actions. Their contribution is helpful to understand how complex a capacity building action can be if it purports to address all the multiple, internal and external, aspects that have an influence on the efforts of a TTO and on its real ability to create value out of research results. The chapter is an indispensable reading to then understand some of the choices that in practice have been made by the PROGRESS-TT consortium in delivering the capacity building action.

Fabiola Bertolotti, Elisa Mattarelli and Paula Ungureanu bring the academic perspective of organizational studies in the topic of capacity building in technology transfer. Their contribution reconnects the topic of this book to the main strands of literature in organization theory and organizational behavior that deal with the performance of organizational and inter-organizations units, such as TTOs. This chapter too proves a necessary reading to understand the basic theoretical framework on the constructs that surround the notion of capacity building when organizations and individuals are involved. Importantly, the chapter provides a unique opportunity to understand that TTOs do not work in an "empty cabinet", as the authors say, but in a context that can be extremely volatile, uncertain, complex and ambiguous.

These three chapters lay the ground for the more specific chapters that deal with capacity building and with the purpose to outline a minimum content for actions aimed at improving the capacity of TTOs and other technology intermediaries.

One of the most difficult tasks in devising a capacity building action towards TTOs (PROGRESS-TT, as a pilot, was supposed to assist thirty TTOs over three years, split in two calls) is the kind of support that is expected. Training is typically a one-to-many activity, with the trainer facing an audience and a content that reflects the expertise of the speaker and its knowledge of the field. If the notion of capacity building aims to be something different and not aimed at individuals only, the approach needs to be one that puts the mentor in direct relation with the TTO, in a genuine one-to-one setting.

Even so, an issue arises, since individual actions need some degree of standardization to be both viable and sustainable and, at the same time, consistent. How can this paradigm shift be achieved while ensuring compliance with such other constraints? Here the PROGRESS-TT comes into play. Andrea Basso, Alan Kennedy and Célia Gavaud in their chapter explain the journey and the efforts to identify a common ground of intervention for the mentors that is the core of PROGRESS-TT and a seminal attempt to make capacity building meaningful from a content perspective. Before that, Alan Kennedy and Pete Frederick illustrate another important component of the overall strategy to frame a capacity building action for TTOs. They explain how the CCODE methodology, once created for SMEs, has been adapted and made validly applicable to TTOs.

There appear to be two distinctive outcomes of PROGRESS-TT in the delivery of capacity building actions to TTOs. One is the idea that effectiveness implies focus and improvement can only become possible if the efforts, both of the mentor and of the TTO, concentrate around critical dimensions, as opposed to framing the intervention more disperse. This led to identify bottom up some critical areas of focus (CAFs).

The second relevant outcome is about using case studies to supply the mentor and the TTOs, even beyond PROGRESS-TT, with a variety of useful materials that, clustered around the single CAFs, become the knowledge platform that should ensure consistency in the individual actions. Moreover, away from the one-model-fits-all approach, the wealth of cases studies is aimed at allowing each TTO to identify itself with other experiences, rather than passively benchmarking with superb, but unmatchable, examples.

It is around one of the CAFs—and particularly that concerning access to finance —that Federico Munari and Laura Toschi in their chapter explain the interface between technology transfer and financial channels that can be accessed to support the maturation and the evolution of the technology towards the market. Relying on their previous studies, the authors provide a fresh insight on how universities try to fund the gap between early stage results and the market, and the experience of proof-of-concept and accelerator programs.

A final section of the book hosts a sample of cases studies, selected for critical area of focus by Marcello Torrisi, and a contribution by one of the mentors of PROGRESS-TT (Tom Flanagan) with his mentee (Elke Piessens), that kindly volunteered to make their experience available in delivering capacity building under the new methodology and using the knowledge platform of PROGRESS-TT and its cases studies.

For the sake of clarity, by no means the section with case studies has the ambition to be complete and exhaustive; it is only there to give tangible example of an experiment that so far seems to be promising in giving content to capacity building and lying the premises for a European way of technology transfer that eventually goes beyond the usual narratives around TTOs and they (in-)capacity to be as performant as their non-European colleagues usually are.

Needless to say, it is not a single book that can deal exhaustively with all the many aspects related to capacity building in technology transfer. This is just one first step towards codifying knowledge and experiences and to try to identify the scientific background for capacity building. Although this book cannot—and it humbly declines to—be considered part of the foreground of PROGRESS-TT, it nevertheless draws from many contributions and efforts of the project. The editors are grateful to the authors that gently accepted to contribute their chapters, even if sometimes not directly involved in the capacity building action, and to the members of the PROGRESS-TT consortium. Francesca Bonadei and Maria Cristina Acocella at Springer are gratefully acknowledged for their patience and their continued interest for innovative and challenging topics.

Part I
Current Status in Capacity Building

European Commission Initiatives Supporting Technology Transfer

Patrick McCutcheon

Abstract

Since the early 2000's the European Commission has supported technology transfer as a means to facilitate the transfer of knowledge to enhance innovation and the competitiveness of the EU economy. This support has taken a number of forms. Firstly the Commission has convened a number of groups of experts and issued a number of policy statements drawing attention to the importance of improving linkages between public researchers and industry and regulations and guidelines facilitating knowledge and technology transfer from the former to the latter. Secondly the Commission has funded a number of projects to improve the capacity of public research organisations and higher education institutes performing research to engage in more technology transfer. Finally the Commission has, through its funding programme Horizon 2020 and its rules for participation; respectively created instruments to fund and finance technology transfer and facilitated the claiming of patent costs as eligible costs in its framework research programmes.

1 Political and Policy Framework on Knowledge and Technology Transfer

In various policy statements, the Commission has frequently addressed technology transfer (TT) through the broader lens of knowledge transfer (KT) where the latter is considered to include the former as well as other vectors whereby knowledge is

See https://ec.europa.eu/programmes/horizon2020/.

P. McCutcheon (✉)
DG Research and Innovation, European Commission, Brussels, Belgium
e-mail: patrick.mccutcheon@ec.europa.eu

© Springer International Publishing AG, part of Springer Nature 2019
M. Granieri and A. Basso (eds.), *Capacity Building in Technology Transfer*,
SxI – Springer for Innovation / SxI – Springer per l'Innovazione 14,
https://doi.org/10.1007/978-3-319-91461-9_2

transferred from one sector to the other. The other vectors are mobility of researchers and different forms of collaboration including collaboration agreements as well as contract research and consultancy.

1.1 Expert Groups

In the context of European Research Area activities aimed at implementing the 3% action plan,[1] the Commission convened a group of experts to draw up a set of recommendations regarding the management of intellectual property in publicly-funded research organisations which could serve as a basis for the development of European guidelines. The report of this group identified many of the issues that have been addressed by subsequent initiatives described below.

In their report in 2004[2] the experts identified the processes, good practices and the implications of a more active involvement in the innovation process and interaction with industry and the creation of new companies. The experts reviewed the KT processes and their evolution from an "Open Science" and a "Licensing Model" to the emergence of the "Innovation Model", whereby the Licensing Model, is supplemented by a more active policy of collaborative research with industry, in particular through EC Framework Programmes (FP), and by a pro-active involvement in the creation of spinout companies. One of the main recommendations of the Group was that the adoption of the Innovation Model by European Public Research Organisations (PROs) should be encouraged as the most effective way to produce significant socioeconomic benefits at European level from publicly funded research results.

The report addressed some of the practical issues faced by PROs when nego-tiating collaborative research and consortium agreements with industry and made some recommendations including that mutually acceptable guidelines be developed by common agreement between representative associations of industry and PROs to facilitate and expand collaborative research opportunities. Whereas the group submitted a tentative set of guidelines on collaborative research addressing the issues of ownership, use rights, access to background, management of IPR and compensation is submitted as a starting point, this work was carried further by a specific project DESCA.[3]

DESCA 2020 (Development of a Simplified Consortium Agreement) is a comprehensive Model Consortium Agreement for Horizon 2020. The latest version dates from February 2016.[4] It was initiated by key stakeholders in the EC's 7th Framework Programme for Research and updated for its successor programme Horizon 2020 in consultation with the Research community. DESCA offers a

---

[1]In the context of the EU's Lisbon strategy, the European Council set an objective to raise overall R&D investment to 3% of GDP by 2010.
[2]http://ec.europa.eu/research/era/pdf/iprmanagementguidelines-report.pdf.
[3]http://www.desca-2020.eu/about-desca/what-is-desca/#.
[4]http://www.desca-2020.eu/latest-version-of-desca/desca-2020-version-12/.

reliable frame of reference for project consortia and enjoys broad support within the FP community.

The Group also recommended that the involvement of PROs in the creation of sustainable spinout companies be further encouraged at European level by a number of public policies and support, including downstream support for these companies after they have left the nurturing environment of a PRO. The group was of the view that the missions of the Knowledge Transfer Office (KTO) must be very well defined and the objectives must be realistic and both must be unequivocally endorsed by the PRO management and supported by the researchers and must be communicated and explained to all the parties involved, industry, government and the public.

Whereas the previously described group of experts was convened by the Commission, the European Council in 2003 called for the Open Method of Coordination (OMC)[5] to be applied to research policy under the aegis of the Committee for Scientific and Technical Research (CREST) which regrouped into different themes 25 recommendations of the 3% Action Plan[6] where OMC could be applied. The expert group convened on the topic in its report "*Encourage the reform of public research centres and universities, in particular to promote transfer of knowledge to society and industry*",[7] recommended in its report in 2006 reform of public research centres and universities to promote KT.

Their core message was to integrate demand driven approaches into the planning of research activities as well as into the redefinition of the operational management of the organisation. To enable more efficient contribution to the innovation process, the expert group recommended that universities should see knowledge transfer as an important mission and entrust the management of this set of activities to a professional, well organised and well supported, knowledge transfer unit with a proper knowledge transfer infrastructure and a system of performance assessment.

As KT is not a self-sustaining activity in the early stages it requires dedicated funding. Since successful knowledge transfer cannot be achieved without the cooperation of individual researchers, the expert group recommended that there should be career and financial rewards to motivate researchers engaged in and linked to KT activities.

In addition to deliberations on fostering KT and TT, the Commission also addressed the challenge of evaluating and valuing the actual intellectual assets which are subject of TT transfers.

The starting point was an observation that few firms and public research organisations and higher education institutes performing research systematically

---

[5]OMC is a soft governance tool, agreed between MS at the Lisbon European Council in March 2000 as an instrument for coordinating national policies by collectively defining objectives and indicators in specific areas. The OMC aimed at ensuring satisfactory progress in policy areas that are primarily within MS competence, involving an exchange of information and best practice, fixing European guidelines and translating them into national and regional policies, establishing indicators and benchmarks, periodic monitoring, evaluation and peer review organised as mutual learning processes.

[6]http://ec.europa.eu/invest-in-research/action/2003_actionplan_en.htm.

[7]http://ec.europa.eu/invest-in-research/pdf/download_en/final_crest_report_march2006.pdf.

take stock of their intellectual capital (IC) and of the added value of research. This under-recognition of IC can lead financial markets to favour traditional rather than research-intensive businesses. It also affects the allocation of resources within companies.

In this context, the Commission supported the development of valuation methods by convening an expert group in 2004 which issued a report: "*RICARDIS: Reporting Intellectual Capital to Augment Research, Development and Innovation in SMEs*"[8] on the role of intellectual capital reporting in the research field.

The group identified the relevant categories of intellectual capital and the reasons for which they are important for research intensive SMEs to attract funds from the financial sector and provided comparative analyses based on a number of case studies. They noted that, as the traditional accounting model is based on the principle of *historic cost*, only a very narrow range of intangibles is covered and, as it only provides a record of what has happened in the past, the picture that they provide is incomplete.

Their idea is that an IC Statement would complement a financial statement as it would provide insight into important resources, including knowledge, access to networks and human resources that are not found on the balance sheet. It would also complement a business plan as it shows *how* value will be created through R&D and describes the role of the various components of intellectual capital.

The group noted that tax treatment of Intellectual Capital is governed by the International Accounting Standard IAS 38 which codifies the traditional accounting approach, which defines an asset in such a way as to exclude "assets" that cannot be directly linked to a revenue stream. They recommended that the European Commission support a process of coordination and convergence of guidelines that will empower national policies and will allow translation and adoption in the member states at different speeds and levels and standardisation and diffusion/dissemination of IC Reporting.

A later expert group was convened in 2012 to consider how IP valuation plays a part in the policy for the "Innovation Union" and the bottlenecks that occur. This Group received the mandate to; review valuation methods for IP and their use, to identify bottle necks in the valuation methods used for the purpose of a company's financial reporting, access to finance and litigation, to identify good practices and to recommend possible policy actions.

This group concluded that there is a clear need to increase market actors' confidence and certainty in IP valuation methods as a way to stimulate IP transactions, to support IP based financing and to give companies the tools to provide information about their IP. However, the group also agreed that it is not the lack of valuation methods per se, or even standards for valuing IP that are missing. They identified the challenge that as IP is, by its nature, innovative and therefore different, each case for valuation requires investigation and as a result, IP valuation of a company's assets is an opinion which is valid, at a particular point in time.

The group noted that valuation of IP assets is complicated by the fact that no two IP assets are the same since a requisite for obtaining such rights is that the IP does

---

[8]http://ec.europa.eu/invest-in-research/pdf/download_en/2006-2977_web1.pdf.

not already exist. The very uniqueness of IP makes comparisons with other IP difficult and limits the usefulness of comparison based pricing. As a result, valuations are often based on assumptions about the IP asset's future use, what important milestones will be met and what management decisions will be taken.

In its report in 2013[9] the Expert Group recommended a number of policy actions to increase the efficient use of IP valuation and to make such valuation flexible, transparent and reliable to respond to market requirements. These included:

Establishing a data source containing information on IP transactions.

Creating an organization to oversee IP valuation practice to increase confidence in the quality of valuations being performed and ensure that valuations are in line with generally accepted principles and standards.

Introducing a risk sharing scheme for banks to facilitate IP secured lending to innovative companies, especially SMEs and Introducing an additional reporting section for intangible assets and IP that would increase the transparency of IP value within company accounts.

1.2 Communications from the Commission

1.2.1 Communication on Improving Knowledge Transfer Between Research Institutions and Industry Across Europe

Following the Commission's policy statement in 2006 on a broad-based innovation strategy for the EU,[10] in which the importance of improving knowledge transfer was identified by the Commission as one of ten key areas for action, the Commission issued a Communication in 2007[11] which set out ideas on how Member States and the Community can act together, to overcome some of the obstacles to KT. It was accompanied by a set of "voluntary guidelines for universities and other research institutions to improve their links with industry across Europe" based on good practices identified by a number of national public authorities and various European stakeholder associations.

In this Communication, the Commission stated that the need for sharing knowledge between research institutions and industry had become increasingly evident in recent years in the context of *open innovation* approaches to R&D, where public research is increasingly a strategic resource and conversely that research institutions need to play a more active role in their relationship with industry in order to maximize the use of the research results and this requires specialist staff to identify and manage knowledge resources with business potential, to make it happen, and to obtain adequate buy-in by all stakeholders. The Communication reflected the conclusions of the previous mentioned expert groups and addressed the

---

[9]http://ec.europa.eu/research/innovation-union/pdf/KI-01-14-460-EN-N-IP_valuation_Expert_Group.pdf#view=fit&pagemode=none.
[10]Putting knowledge into practice: A broad based innovation strategy for the EU—COM (2006) 502. http://eur-lex.europa.eu/legal-content/EN/TXT/HTML/?uri=LEGISSUM:i23035&from=EN.
[11]http://ec.europa.eu/invest-in-research/pdf/com2007182_en.pdf.

issues of technology transfer skills and competences, pooling resources and promotion of an entrepreneurial mind-set.

Attached to the Communication was a set of guidelines highlighting good practices regarding the management and transfer of knowledge and intellectual property. The guidelines firstly outlined the issues to be addressed by research institutions to ensure their policies on IPRs, manage incentives and conflict of interest actually optimise knowledge transfer activity. They secondly comprised good practices relating to contractual arrangements which should be taken into account when negotiating research collaboration contracts.

The issues concerned highlighting the benefits to research institutions of engaging in knowledge transfer, striking a balance between openness and exploitation of results, elements of what an adequate IP policy should comprise as well as the need for incentives and resource implications. The good practices addressed i.a. the negotiation process, rights of parties, use of model contracts and IP enforcement.

The reaction to this communication was a call on the Commission to make specific recommendations to the Member States and this was done through a follow-up Commission Recommendation[12] in 2008. This recommendation comprised 11 specific recommendations The Communication was accompanied by a code of practice with specific recommendations to PROs on intellectual property and KT policies and examples of good practice on i.a. managing intellectual property, capacity and skills and was warmly endorsed by the Council.[13]

1.2.2 Communication on the European Research Area

The European Research Area which comprises the various stakeholders engaged in research and innovation and their various interactions has also addressed knowledge transfer. In its most recent policy statement in 2012,[14] the Commission committed itself to carrying out an assessment of existing initiatives and developing a comprehensive policy approach to open innovation (OI) and knowledge transfer (KT).

Following this a group of experts convened by the Commission noted in 2014[15] that Europe faces existential challenges on how to create sustainable growth given the vast overhang of public and private debt and how to do this given the transformational impact of disruptive technologies on traditional models for business and public sector organizations, banks, universities and public research organizations (PROs).

Their report sets out to develop coherent whole of policy recommendations for OI and KT, across four priority areas. The experts concluded that placing OI and KT in the spotlight requires an Open Innovation 2.0 policy, a harmonized European high quality, informed and influential IP policy and that embracing innovative businesses and growing innovative markets should be a core aim of the EC and

---

[12]http://ec.europa.eu/invest-in-research/pdf/ip_recommendation_en.pdf.
[13]http://ec.europa.eu/invest-in-research/pdf/download_en/st10323_en08.pdf.
[14]http://ec.europa.eu/research/era/pdf/era-communication/era-communication_en.pdf.
[15]http://ec.europa.eu/research/innovation-union/pdf/b1_studies-b5_web-publication_mainreport-kt_oi.pdf#view=fit&pagemode=none.

making Universities and PROs more entrepreneurial by requiring more emphasis on delivery of outputs.

Finally, the experts were of the view that private capital is needed both to effectively address the financing and funding gaps in proof of concept projects, the early to mid-stage development of start-up companies and the growth phase of promising SMEs in Europe.

1.2.3 Communication on Innovation Union

In 2010, in the framework of the Lisbon Strategy[16] the Commission issued its most recent policy statement on innovation[17] and the first of which received the support of heads of government. It includes a series of actions to promote KT.

Among the 34 commitments, the Commission undertook to make proposals to develop a European knowledge market for patents and licensing. Aided by a group of experts, The Commission reviewed existing experiences in trading platforms that match supply and demand, market places to enable financial investments in intangible assets, and other ideas for breathing new life into neglected intellectual property, such as patent pools and innovation brokering. It concluded[18] that the main line of intervention at EU level should focus on maturing the technologies rather than on bundling the patents that read upon the technologies. On forms of aggregation and intermediation it noted that a plethora of business models were emerging and that in general there was no need for an EU level intervention.

The Commission also undertook examine the role of Competition Policy in safeguarding against the use of intellectual property rights for anti-competitive purposes. It did so in its review and revision of the block exemption regulations for R&D agreements in 2010[19] and technology transfer agreements[20] in 2014 along with a revision of the related guidelines of the latter. These regulations and guidelines explain how such agreements can be constructed in compliance with and not fall foul of Arts 101 and 102 TFEU.[21] As all block exemption regulations in the area of competition policy, these regulations and their related guidelines are subject to periodic review and where warranted revision.

---

[16]http://www.europarl.europa.eu/summits/lis1_en.htm.
[17]http://ec.europa.eu/research/innovation-union/pdf/innovation-union-communication_en.pdf#view=fit&pagemode=none.
[18]http://ec.europa.eu/research/innovation-union/pdf/expert-groups/report_of_the_expert_group_on_patent_aggregation_-_2015.pdf#view=fit&pagemode=none.
[19]http://eur-lex.europa.eu/LexUriServ/LexUriServ.do?uri=OJ:L:2010:335:0036:0042:EN:PDF.
[20]http://eur-lex.europa.eu/legal-content/EN/TXT/PDF/?uri=CELEX:32014R0316&from=EN.
[21]http://eur-lex.europa.eu/legal-content/EN/TXT/PDF/?uri=OJ:C:2008:115:FULL&from=EN.

2 European Commission Funded Projects and Studies

Following the recommendations of experts and its own policy statements as described in Sect. 1 see above, the Commission has funded a number of projects all of which aim to raise the capacity of research teams and SMEs to engage in technology transfer. These projects focused on variously, strengthening networks of technology transfer offices as well as the offices themselves, providing frontline advice, capacity building aimed at facilitating access to financial instruments and in commercialising the results of EU funded projects.

2.1 ProTon Europe

ProTon Europe started as an innovation support project funded by the Commission following a call in 1999 for proposals to create a pan-European representative association of KT offices from PROs.

The aim of the project was to create a self-sustainable[22] pan-European representative association of knowledge transfer offices from public research institutions. The specific objectives were to facilitate networking among public research organisations transfer offices, exchange of experience and good practice at European level, contribute to raising general professional standards, provide an extensive geographical scope to networking activities (including regions and operators that may usefully benefit from the interaction (in particular regions of the EU and Associated States where this activity was hitherto less developed)) and provide Commission services with access to these operators, in order to extract innovation policy lessons at European level.

ProTon Europe provided input to the Commission on behalf of its members and partner associations on the revision of the State Aid Framework for Research and Innovation, which was adopted at the end of 2006. ProTon Europe also provided input to the Commission on patent policy. The European Institute of Technology as well as CREST working groups on IP matters and training and University–Industry collaboration. In collaboration with the European University Association (EUA),[23] with the European Industrial Research Management Association (EIRMA)[24] and with the European Association of Research and Technology Organizations (EARTO)[25], ProTon pursued the promotion of the Responsible Partnering initiative in collaborative research, launched at the European Business Summit in March 2006 and revised in 2009.[26] The partners are currently engaged in a process of further updating the initiative.

---

[22]Self-sustainable in this context means sustainable from a financial perspective, i.e. financial autonomy from the EC at the end of the project.
[23]http://www.eua.be/.
[24]http://www.eirma.org/.
[25]http://www.earto.eu/.
[26]http://www.eua.be/Libraries/publications-homepage-list/Responsible_Partnering_Guidelines_09.pdf?sfvrsn=2.

Until it merged with ASTP in 2014,[27] ProTon continued as an independent organisation after the project and continued to provide services providing access to the policies and processes by which members have solved common issues; establishing staff mobility schemes; and bringing people together to discuss and resolve new, and long term, shared issues and running an Annual Survey to facilitate both measurement of knowledge transfer activity, self-assessment by individual KT Offices and international benchmarking.

Post project ProTon continued to represent the profession of knowledge transfer at European level, complement the work of existing national associations and support the creation of new ones, contribute to the development and recognition of the knowledge transfer function within public research organisations by proposing good practices, ethical and professional standards of conduct and of reporting as well as to boost knowledge transfer by providing a comprehensive range of services and benefits to its members, including training, networking, exchange of experience and facilitating trans-European knowledge transfer.

Compared with ASTP with which it merged, ProTon's membership was open to organisations (not individual persons) involved in the transfer of knowledge from PROs with a substantial activity in the transfer of knowledge, including intellectual property management, licensing, partnering with industry and the creation of new companies.

2.2 The EU IPR Helpdesk

The IPR-Helpdesk started[28] as a project funded by the European Commission and was launched in 1998 as a central reference point for intellectual property inquiries and advice throughout the European Union. The IPR-Helpdesk is implemented by a European network consisting of several research institutes, law firms and consultancies. It offers a free-of-charge enquiry service, or Helpline service, for addressing intellectual property issues, that is targeted at researchers and European SMEs participating in EU-funded collaborative research projects.

The European IPR Helpdesk[29] is now a project funded by the European Commission under the current Horizon 2020 programme that supports cross-border SME and research activities to manage, disseminate and valorise technologies and other Intellectual Property (IP) Rights and IP assets at an EU level.

It offers a broad range of informative material, a Helpline service for direct IP support and on-site and online training The IPR Helpdesk's main goal is to support IP capacity building from awareness to strategic use and successful exploitation. This

---

[27]ASTP, the Association of Science and Technology Professionals was founded in 2000 and was a similar organisation whose mission was to professionalize and promote technology transfer between the European science base and industry by organising conferences and training courses and master classes to improve technology transfer skills for all levels of experience and provide a virtual community for year round exchange of best practices by its members.

[28]http://cordis.europa.eu/news/rcn/11447_en.html.

[29]https://www.iprhelpdesk.eu/.

strengthening of IP competencies focuses on EU SMEs, participants and candidates in EU-funded projects, and EU innovation stakeholders for an increased translation of IP into the EU innovation ecosystem. The Helpline offers support on specific IP issues from a team of experienced legal specialists within no more than three working days. In the provision of its advice the IPR Helpdesk treats all information and documents received and dealt with by the Helpline in the strictest confidence and commits to not make use of, nor divulge any information to third parties.

Beyond specific queries, the Helpdesk also provides, on subscription, a European IPR Helpdesk email newsletter and Bulletin to keep track on latest developments in the field of IP and a series[30] of online fact sheets, case studies, guides and informative materials as well as free webinars, or meet the team at training events and conferences.

In addition to the EU IPR helpdesk, the Commission has set up other IPR helpdesks[31] to assist SMEs with IPR issues when internationalising their activities in ASEAN, China and Latin America. These Helpdesks supports EU SMEs to both protect and enforce their IP rights in or relating to China, Latin America and, Southeast Asia through the provision of free information and services. These take the form of first-line, confidential advice on intellectual property and related issues, plus training, materials and online resources.

2.3 EUKTS European Knowledge Transfer Society

The EUKTS project[32] was set up to increase the standard and recognition of the knowledge and technology transfer (KTT) profession across Europe. It is now running as an independent organisation[33] with a secretariat in Brussels.

The project specifically addressed the OMC CREST recommendation regarding the need for a set of training and professional quality standards for KTT practitioners and European Commission policy statements that had identified the area of KTT as a structural weakness. The challenge related to the attractiveness of KTT as a career, the absence of standardised training and the lack of KTT certification with Europe-wide recognition. The EUKTS project was based on the experiences of the certified transnational technology transfer manager project—CERT-TTT-M[34] in which a training framework for practitioners in KTT field had been defined.

The principal aim of the EUKTS was to establish a European not-for profit organisation with the main functions of providing accreditation for KT training programmes, managing the process of certification of KT professionals, and providing international recognition for KT professionals. The project aimed to increase the status and recognition of the KTT profession by developing adequate professional and training standards including the definition of the professional KTT

---

[30]https://www.iprhelpdesk.eu/library.
[31]http://www.ipr-hub.eu/.
[32]http://cordis.europa.eu/result/rcn/54403_en.html.
[33]https://www.eukts.eu/.
[34]http://www.ttt-manager.eu.

curriculum and procedures for accreditation of courses and certification of individuals at different stages of their professional career. The recognition process was structured on procedures for accreditation and reaccreditation of courses and certification and recertification of KTT professionals.

During the project, the consortium, defined the aims, functions of the future EUKTS association and set out a framework for certification along with a model of curriculum of skills and competencies for the KTT profession. The EUKTS curriculum covers the following core competences; information analysis and management, Intellectual property (IP) protection, strategy and management of IP portfolios, opportunity assessment and valuation, KT and marketing of innovations, negotiating and contracting, new business development and financing and project management.

The scheme was supported by a feasibility study providing details on internal organisational and legal aspects including the economical sustainability. It was successfully tested in Austria and the Czech Republic. This pilot phase enabled the implementation and fine tuning of the certification scheme of KTT professionals from all sectors including industry, universities, other public research organisations, and government.

EUKTS is now structured as an association with members representing all types of stakeholders in the value chain of KTT; professional KT organisations based in the EC, industry associations, and regional agencies for innovation, the IP Offices, and industry professional groups involved in technology transfer.

EUKTS accreditation of training courses is designed to provide assurance on the quality and relevance of the course content to KTT practitioners and the quality assurance measures and procedures adopted by the provider in delivering the courses. Accreditation can be requested by all types of training providers: universities, private providers, groups of providers. Four programmes are in the process of accreditation or reaccreditation.

EUKTS professional certification has been developed to recognise KTT professionals at various stages in their careers and is offered at three levels: EuKTS Associate for junior professionals, EUKTS Professional for practitioners with a minimum 5/7 years of experience, EuKTS Expert for advanced experts. The scheme includes procedures for grandfathering which has been launched in June 2016. Within the last months, already more than 50 certifications were granted.

The EUKTS Certification Scheme is being continuously upgraded with the stakeholders in both the field of KTT and Certification[35] and, with the support of members and key partners including IP Offices. The next steps of development are now to disseminate the EuKTS scheme and digitise the procedures and exam content to enable access to certification from everywhere. Thus, the EuKTS is continuing to contribute to the improvement of private/public cooperation, increase the capability of evaluating and exploiting research results, leading to greater efficiencies and competitiveness in of KTT activities in Europe.

---

[35]www.efcocert.eu.

2.4   Knowledge/Technology Transfer Indicators

In the context of ERAC, the Commission convened an expert group to examine the case for the recognition of a headline indicator as well as consider elements that could be used to develop a composite knowledge transfer indicator. On the first question the experts highlighted in a report in 2010 the advantages and drawbacks of two possible headline indicators.[36] On the elements of the composite indicator which considered not only technology transfer but also collaborative research and mobility of scientists, the group developed in 2011 a concept for such an indicator.[37] The challenge with implementing such an indicator was the actual availability of data on all of the indicators proposed. Nevertheless the work of this group has since been used by ASTP-PROTON in the development of its surveys of its members' technology transfer performance.

2.5   More Recent Commission Projects

2.5.1   Progress TT

In anticipation of the launch of the InnovFin financial instrument on TT[38] in 2016, the Commission decided to facilitate access to this instrument by providing capacity building to TTOs to improve their investment readiness. Following a call for proposals, the Commission awarded a contract to a consortium which proposed a series of actions as follows.

The Commission in developing this TT instrument was conscious that such an instrument might merely facilitate more TT transactions by those already active. In order to encourage participation by HEIs and PROs who are not yet known to the investment community, Progress TT aimed to raise the investment readiness of TTOs not already accessing private finance.

This project, on the basis of a self-assessment tool, categorised TTOs according to their capacity and offered support to those in the second and third tiers and identified best practices for a number of areas of critical focus bases on a survey of needs and leveraged these to build training materials. The support involved trainings, workshops, boot-camps and one-on-one mentoring support.

This project ended in December 2017 and it is foreseen that it will be followed by a more direct service provision of IP advice.

2.5.2   Entente Health

In parallel, in the field of medicine research, the Commission financed a project[39] Entente Health which aims at strengthening KTOs in universities, public research organisations, and hospitals and at promoting transnational collaboration between

---

[36]http://ec.europa.eu/research/innovation-union/pdf/kt_headline_indicators.pdf#view=fit&pagemode= none.

[37]http://ec.europa.eu/research/innovation-union/pdf/kti-report-final.pdf#view=fit&pagemode= none.

[38]http://eif.europa.eu/what_we_do/equity/single_eu_equity_instrument/innovation_equity/index.htm

[39]http://entente-health.eu/.

industry and academia in the health sector, through networking activities among all the key stakeholders within knowledge transfer in the health sector in Europe. It does so by providing the following services; guidance on different aspects of technology transfer, a library of best practices, success stories and policy measures, concerning IP assets, KT activities and types of agreements as well as the possibility to participate in TTO exchange. The aim of this latter is to develop the experience and expertise of technology transfer professionals by working alongside seasoned TTOs, industry or investment practitioners to learn by direct experience to develop a clearer understanding of the processes, priorities and overall business culture.

2.5.3 Common Exploitation Booster

Grant beneficiaries under Horizon 2020 could avail of a support service, the Common Exploitation Booster[40] which aimed to bridge the gap between research results and exploitation by helping the project consortia in raising awareness on exploitation possibilities and exploitation planning, clarifying issues, exploring solutions and actions, anticipating possible conflicts for successful exploitation, setting up roadmaps for the long-term sustainability of the project results and creating value out of novel knowledge.

This initiative provided; analysis of exploitation risks to scout the route towards the market and better tackle risks, exploitation strategy seminars involving joint working sessions to streamline the exploitation strategy and a go-to-market action plan, business plan development to design a convincing and actionable plan for exploitation as well as brokerage and pitching events where partners present their results to peers, potential users and investors.

2.6 JRC TTO Circle

In addition to the projects described above, the Commission's Joint Research Centre has established a network of TTOs, the TTO circle,[41] which membership is open to TTOs of European public research organisations. The TTO circle organises workshops and events to explore solutions to emerging issues facing the TTO community as well as some training which it carries out in cooperation with WIPO.

3 Financial Support

3.1 IPR Cost as Eligible Costs

Under the rules for participation in Horizon 2020 per Regulation (EU) No. 1290/2013[42] projects may contribute to a maximum of 100% of the total

---

[40]http://exploitation.meta-group.com/Documents/short_leaflet.pdf.
[41]https://ec.europa.eu/jrc/communities/community/tto-circle-community.
[42]http://www.fch.europa.eu/sites/default/files/h2020-rules-participation_en.pdf.

costs of activities concerning protection, dissemination and management of IPR. Protection of IPR falls under the normal rule of eligible costs per Article 28. That protection is an eligible cost is explicitly laid down in Article 6.2.D.3 of the Horizon 2020 Model Grant Agreement[43] which explicitly provides that:

> Costs of other goods and services (including related duties, taxes and charges such as non-deductible value added tax (VAT) paid by the beneficiary) are eligible, if they are (a) Purchased specifically for the action and in accordance with Article 10.1.1 or (b) Contributed in kind against payment and in accordance with Article 11.1.

> Such goods and services include, for instance, consumables and supplies, dissemination (including open access), **protection of results**, certificates on the financial statements (if they are required by the Agreement), certificates on the methodology, translations and publications.

This is further clarified in the Annotated Model Grant Agreement (MGA):

> Costs related to the protection of the actions results (e.g. consulting fees, fees paid to the patent office for patent registration; see Article 27) are eligible if the eligibility conditions are fulfilled. Costs related to protection of other intellectual property (e.g. background patents) are NOT eligible.[44]

> Costs related to protection may be eligible (see Article 6.2.D.3).[45]

The basic requirement is that such costs are incurred during the action, are necessary for the implementation of the action and are reasonable, justified and comply with the principle of sound financial management, in particular regarding economy and efficiency.

Royalties paid for IPR access rights (and by extension any lump sum payments) are normally eligible, if all the above mentioned eligibility conditions are met.

However the following royalties are however **NOT** eligible (or eligible only within certain limits):

Royalties for an exclusive licence are eligible only if it can be shown that the exclusivity is absolutely necessary for the implementation of the action.

In the case of royalties for licensing agreements which were already in force before the start of the action, only the part of the licence fee that can be linked to the action is eligible.

Royalties for access rights to background granted by other beneficiaries are not normally eligible.

The rationale for the latter is that since the default rule is that access rights are granted on a royalty-free basis, beneficiaries may deviate only if agreed before GA signature and royalties are eligible only if explicitly agreed by all beneficiaries before GA signature and all the other eligibility conditions are met. If beneficiaries intend to deviate from the default rule, it is recommended that this is explained in detail in their proposal.

---

[43]http://ec.europa.eu/research/participants/data/ref/h2020/grants_manual/amga/h2020-amga_en.pdf.
[44]Ibid., page 89.
[45]Ibid., page 227.

Costs related to the protection of the actions results are eligible if the eligibility conditions are fulfilled. Costs related to protection of other intellectual property (e.g. *background patents*) are generally NOT eligible. Costs for drawing up the plan for the exploitation and dissemination of the results are normally NOT eligible since they will have been incurred before the start of the action, to prepare the proposal. Costs that occur when revising or implementing this plan may be eligible.

In addition to allowing some IP costs as eligible costs the Rules for Participation encourage the protection and exploitation of the results of the funded project per Art 42 and 43 of the Rules for Participation. The beneficiaries must for any results that can reasonably be expected to be commercially or industrially exploited: examine the possibility of protecting them and if possible, reasonable and justified, protect them even if this requires further research and development or private investment.

The beneficiaries are in principle free to choose any available form of protection.

The choice of the most suitable form should be made on the basis of the specificities of the action and the type of result (i.e. the form which offers the most adequate and effective protection). Although important for commercial and industrial exploitation, IP protection is not mandatory.

3.2 Financial Support for Technology Transfer

3.2.1 Grant Schemes SME Instrument, Fast Track to Innovation and Eurostars

Under Horizon 2020 Innovative SMEs have access to grants that facilitate the commercialisation of their research projects and or to innovate on the basis of research carried out by other organisations. The SME instrument[46] is grant based and offers grants to individual beneficiaries in two phases. On completion of the two phases, projects should be investment ready and capable of attracting private finance from i.e. the accredited intermediaries of the EIF which can offer debt and equity finance under InnovFin, see below.

Horizon 2020 funds high-potential innovation through a dedicated SME instrument, which offers seamless business innovation support under the section Societal Challenges and the specific part Leadership in Enabling and Industrial Technologies (LEITs).

With €3 billion in funding over the period 2014–2020, the SME Instrument is aimed at high-potential SMEs to develop groundbreaking innovative ideas for products, services or processes that are ready to face global market competition. Available to SMEs only,[47] the scheme offers a highway to innovation through phased, progressive and complimentary support. Specifically it offers Business innovation grants up to €50,000 for feasibility assessment purposes (phase I), Business innovation grants for innovation development & demonstration purposes

---

[46]http://ec.europa.eu/programmes/horizon2020/en/h2020-section/sme-instrument.

[47]SMEs can however organise a project in the way that best fits their business needs—meaning that subcontracting is not excluded.

(phase II) of between €0.5 m and €2.5 m and a free-of-charge business coaching to support and enhance the firm's innovation capacity and help align the project to strategic business needs.

Activities funded in phase I could include; risk assessment, design or market studies, intellectual property exploration; the ultimate goal is to put a new product, service or process in the market, possibly through an innovative application of existing technologies, methodologies, or business processes. Activities funded in phase 2 can include; prototyping, miniaturisation, scaling-up, design, performance verification, testing, demonstration, development of pilot lines, validation for market replication, including other activities aimed at bringing innovation to investment readiness and maturity for market take-up.

The Fast Track to Innovation pilot[48] is a similar programme based on grants to consortia rather than single beneficiaries. It is a fully-bottom-up measure in Horizon 2020 promoting close-to-the-market innovation activities that is open to all types of participants. It aims to reduce the time from idea to market and to increase the participation in Horizon 2020 of industry, SMEs and first-time industry applicants and to nurture trans-disciplinary and cross-sector approaches. It also covers sustainable innovations addressing societal needs or areas under LEITs and, at the same time, creates viable business opportunities. On offer is a maximum EU contribution of €3 m per proposal.

As with traditional EU R&D programmes, proposals for funding must be submitted by consortia comprising between three and five legal entities established in at least three different EU Member States or countries associated to Horizon 2020. Within each consortium there must either be an allocation of at least 60% of the budget to industry participants or the consortium must include a minimum of two industry participants in a consortium of three or four partners, or three industry participants in a consortium of five partners.

The Eurostars Programme[49] is a joint programme between EUREKA[50] and the European Commission dedicated to research-performing SMEs. Eurostars stimulates them to lead international collaborative research and innovation projects by easing access to support and funding. The Programme is a Programme co-funded by the European Communities and 36 EUREKA member countries. It aims to stimulate these SMEs to lead international collaborative research and innovation projects by easing access to support and funding. It specifically targets the development of new products, processes and services and the access to transnational and international markets.

Based on Article 185 of the Lisbon Treaty, Eurostars aims to combine with a bottom-up approach, a central submission and evaluation process, and synchronized national funding in the 36 countries. Eurostars projects must involve at least two participants (legal entities) from two different Eurostars participating countries and

---

[48]https://ec.europa.eu/easme/en/fast-track-innovation-fti-pilot.
[49]http://www.eurekanetwork.org/eureka-eurostars.
[50]EUREKA is an intergovernmental network supporting market oriented RD&I projects, facilitating innovation and access to finance http://www.eurekanetwork.org/.

the main participant must be a research-performing SME from one of these countries.

The role of the SME participants in the project should be significant. At least 50% of the project's core activity should be carried out by SMEs. The consortium should be well balanced, which means that no participant or country can have more than 75% of the budget of the total project costs.

A Eurostars project should be market-driven: it must have a maximum duration of three years, and within two years of project completion, the product of the research should be ready for launch onto the market. The exception to this rule applies to biomedical or medical projects, where clinical trials must be started within two years of project completion.

These activities now comprise the European Innovation Council (EIC) pilot.[51]

3.2.2 InnovFin Equity Technology Transfer

Whereas spinoffs from public research organisations and higher education institutes performing research constitute SMEs and can have access to the above mentioned scheme, projects that are at a pre-incorporation do not have access to the above instruments. For this reason a specific instrument on technology transfer[52] has been developed under InnovFin Equity[53] to provide equity financing for the commercialisation of the results of public funded and performed research.

InnovFin Equity, which is managed by EIF, is part of "InnovFin[54]—EU Finance for Innovators", an initiative launched by the European Commission and the EIB Group in the framework of Horizon 2020. InnovFin Technology Transfer (TT), part of the InnovFin Equity umbrella, targets investments into technology transfer funds operating in the pre-seed (including proof of concept) and seed stages, in particular funds affiliated or cooperating with Technology transfer offices, Research organisations, Universities and higher education institutes (HEIs) performing research and TT specialists.

InnovFin TT aims to accelerate technological innovations, especially in the areas of key enabling technologies and other Horizon 2020 objectives, by maturing technologies, including the promotion of IP by assigning of or licensing-out IPRs and supporting spin off activities. InnovFin TT is designed to support the commercialisation of research results through proof of concept, project (POC) financing, and/or commercialisation, IP exploitation (licensing, sale of patents) or spin-outs, spin-offs or joint venture activities.

The pre-seed (proof of concept) stage involves the pre-incorporation phase with feasibility testing, technology/product/process for commercialisation, prototyping and incubation. This phase mainly comprises products and technologies with a Technology Readiness Level (TRL) between TRL 3 to TRL 6 or the equivalent Innovation Readiness Level (IRL) maturity between IRL 1 and IRL 2. The project

---

[51]https://ec.europa.eu/research/eic/
[52]http://www.eif.europa.eu/news_centre/publications/innovfin_technology_transfer.pdf.
[53]http://www.eif.europa.eu/what_we_do/equity/single_eu_equity_instrument/innovfin-equity/index.htm.
[54]http://www.eib.org/products/blending/innovfin/.

financing stage follows the proof of concept and is foreseen in life science projects at both pre-clinical and clinical stages. The seed stage includes demonstration, low scale production or development of associated business applications of tested and validated products/technologies. This phase mainly comprises products and technologies with a TRL between TRL 7 to TRL 8 or the equivalent IRL maturity between IRL 3 and IRL 4.

InnovFin TT investments are carried out on a pari-passu basis with other investors in the fund. The fund's lifetime may not exceed 20 years and the fund shall have at least 30% of its total commitments coming from private investors and the fund's manager must be established or operating in one of the Participating Countries.

Finance is made available via accredited intermediaries and beneficiaries may request as an optional initial step POC funding. Following this the route to commercialisation may take either the route of incorporation of a spin-off or project finance whereby the further development will be financed with the aim of licensing the results.

3.2.3   ERC Proof of Concept and EIT Impact Fund

The European Research Council (ERC)[55] which provides grants to single beneficiaries has also established a POC fund[56] which is limited to the beneficiaries of ERC grants. All Principal Investigators in an ERC frontier research project, that is either on going or has ended less than 12 months before 1 January 2017, are eligible to participate and apply for an ERC POC Grant. The Principal Investigator must be able to demonstrate the relation between the idea to be taken to proof of concept and the ERC frontier research project in question.

The ERC POC funding is made available to those who already have an ERC award to establish proof of concept of an idea that was generated in the course of their ERC-funded projects. The activities to be funded shall draw substantially on the ERC-funded research. However the additional funding is not aimed at extending the original research but is predominantly concerned with overcoming obstacles to practical application.

The funding will cover activities at the very early stage of turning research outputs into a commercial or socially valuable proposition, i.e. the initial steps of pre-competitive development. The funding can be used to; establish viability, technical issues and overall direction, clarify intellectual property rights positions and strategy, provide feedback for budgeting and other forms of commercial discussion, provide connections to later stage funding and cover initial expenses for establishing a company.

The European Institute of Technology[57] is reflecting on the possibility of establishing an innovation impact fund to commercialise the outcome of its knowledge and innovation communities (KICs).

---

[55]https://erc.europa.eu/.
[56]https://erc.europa.eu/funding/proof-concept.
[57]https://eit.europa.eu/.

3.2.4 EU Policies Financed Either Fully or Partially by Member States

In addition to the above described schemes funded under Horizon 2020, mention should be made of the EU RD&I framework and EU cohesion policy. The former sets out the rules by which member states may subsidise RD&I activities and the latter comprises co-funding by the Commission and Member States.

EU RDI State aid framework

The EU RDI State aid framework[58] serves to clarify the State aid rules applicable to Research, Development and Innovation (RD&I) funding activities. This includes the financing of knowledge transfer activities. The framework explains how RD&I activities can comply with and not fall foul of the Treaty provisions against State aid per Art 107 and 108 TFEU which aim is to ensure compatibility with the internal market.

Knowledge transfer activities have a non-economic character if they are "internal" and all income from these activities is reinvested in the primary activities of the research organisations. In contrast, any economic activity performed should take place under normal market conditions, and public funding of such activities will generally be considered to constitute State aid and needs to comply. The framework also provides that research institutions should separately allocate costs and revenues to economic and non-economic activities, in order to avoid possible cross-subsidisation.

EU cohesion policy

Support for knowledge transfer activities is available through EU cohesion policy. Cohesion policy's main instrument, the European Regional Development Fund (ERDF),[59] is used to support incubators and science parks (infrastructures and accompanying services) which are an effective means to spin-out knowledge into the market place and can help create better SMEs—university links.

Furthermore, the European Social Fund (ESF) provides financial support through the assistance to persons (training, guidance, etc.) and for the development and modernisation of educational structures and systems. Since 2007 there has been an increased emphasis on strengthening research and innovation, particularly through knowledge transfer.

Observations

As outlined in this chapter, the Commission has undertaken a considerable amounting of consulting with experts and financed studies on different aspects

---

[58]http://eur-lex.europa.eu/legal-content/EN/TXT/PDF/?uri=CELEX:52014XC0627(01)&from= EN.
[59]http://ec.europa.eu/regional_policy/en/funding/erdf/.

related to technology transfer in the context of its innovation policy development. This has led to a number of policy statements with recommendations for action by different stakeholders as well as development of a number of financial instruments and grant support schemes as well as capacity building activities. At the time of writing the Commission is preparing the elements of the next multi-annual programme. Although this process is a work in progress, based on the initial outcomes of the initiatives described a number of developments are foreseeable.

These include the development of further financial instruments focused on countries classified as moderate innovators according to the Innovation Scoreboard,[60] revision of the SME instrument as a bottom-up instrument in the context of the new European Innovation Council,[61] one-on-one support in relation to IP to SMEs and TTOs which go beyond what is offered by the IPR helpdesks and is available as eligible costs under Horizon 2020. On the policy side, the competition rules related to State aid the RD&I framework and the block exemption regulations on R&D agreements and related guidelines will in due course be reviewed and possibly further revised.

---

[60]http://ec.europa.eu/growth/industry/innovation/facts-figures/scoreboards_en.
[61]https://ec.europa.eu/research/eic/index.cfm.

Intervention Opportunities for Capacity Building in Technology Transfer

Lutz Maicher, Katja Dralle Mjos and Liina Tonisson

Abstract

Within the last decades, an extensive body of literature has focused on the question how technology transfer effects the returns on investments in the research system, on the university level and the regional or national level alike. In conjunction with this research in innovation economy, the question on success factors of technology transfer offices and activities arose. Why are certain universities or regions more successful in technology transfer than others? In the United States the revenues from technology transfer have increased almost tenfold from $160 million in 1991 to $1.4 billion in 2005, reaching $2.6 billion in 2012 (AUTM 2013). However, this success appears to be highly volatile. Some very successful universities earn a big proportion from this budget, while the majority of smaller technology transfer offices are struggling with earning their expenses. On a lower magnitude, this effect can be observed in Europe as well. Previous research showed that there is no clear evidence for this difference in performance, which rather appears to be the result of an intertwined web of various influence factors. Knowing the determinants on the success of technology transfer offices is important for the design of any kind of capacity building programs, as these influence factors present potential intervention opportunities for improving the capacity of technology transfer offices. This

L. Maicher (✉)
Jena University and Fraunhofer IMW, Leipzig, Germany
e-mail: lutz.maicher@imw.fraunhofer.de; lutz.maicher@uni-jena.de

K. D. Mjos
Fraunhofer Center for International Management and Knowledge Economy,
Leipzig, Germany
e-mail: katja.mjos@uilo.ubc.ca

L. Tonisson
TROPOS, Leibniz Institute for Tropospheric Research, Leipzig, Germany
e-mail: liina.tonisson@tropos.de

© Springer International Publishing AG, part of Springer Nature 2019
M. Granieri and A. Basso (eds.), *Capacity Building in Technology Transfer*,
SxI – Springer for Innovation / SxI – Springer per l'Innovazione 14,
https://doi.org/10.1007/978-3-319-91461-9_3

chapter provides an in-depth review of the existing research in technology transfer with the goal of spotting the identified influence factors of performance success.

Abbreviations

HR	Human Resources
IP	Intellectual Property
PRO	Public Research Institution
R&D	Research & Development
SME	Small and Medium Enterprise
TT	Technology Transfer
TTO	Technology Transfer Office
VC	Venture Capital

1 Introduction

Technology transfer (TT) helps develop early stage ideas that may have been intellectual property (IP) protected towards a viable product for direct use by the research community. Furthermore, it aims to develop intellectual property into bases for new platforms, mechanisms, or services to be made into goods for public use.

Successful transfer and development of the technology helps promote the public research organization (PRO) and its industry partners (Anderson 2001). Industry can in many cases save on technology development costs, and universities gain recognition and prestige while co-operating with industry for technology development (Segatto-Mendes and Mendes 2006). Moreover, successful technology transfer can support educational and further research activities at universities via licensing revenues (Wayne 2010).

For successful technology transfer, company scientists and university researchers are successfully linked across different universities or industries in order to advance the knowledge in a particular field or to further develop a technology or product. These collaborations may result in licensing or contract research opportunities that benefit both universities and industry (Etzkowitz et al. 1998). In addition, technology transfer ensures that the interests and rights of the university technology transfer office (TTO) with regards to the intellectual property are protected. The TTO is able to retain the intellectual property rights of the technology and issue a license for the conditional use of the technology to industry or another research institute (Powers 2003).

Historically, the US has been at the frontier of successful technology transfer and the difference in university licensing fees between the US and Europe is enormous (AUTM 2013). Based on AUTM (2013) data, the most successful US universities earn a big proportion from overall PRO licensing budget, while the majority of smaller TTOs are struggling with earning their expenses. On a lower magnitude, this effect can be observed in Europe as well; however, there is no clear evidence for this difference in performance but a multitude of influential factors (Etzkowitz et al. 1998). Knowing these determinants on the success of TTOs is vital for the design of any kind of capacity building programs, as these influence factors present potential intervention opportunities for improving the capacity of TTOs.

We have reviewed the state-of-the-art of technology transfer and identified various influence factors of TTO performance success, which we grouped into five fundamental pillars falling within the scope of different stakeholders (Fig. 1). In the following sections each of these five pillars will be discussed in detail, as each of

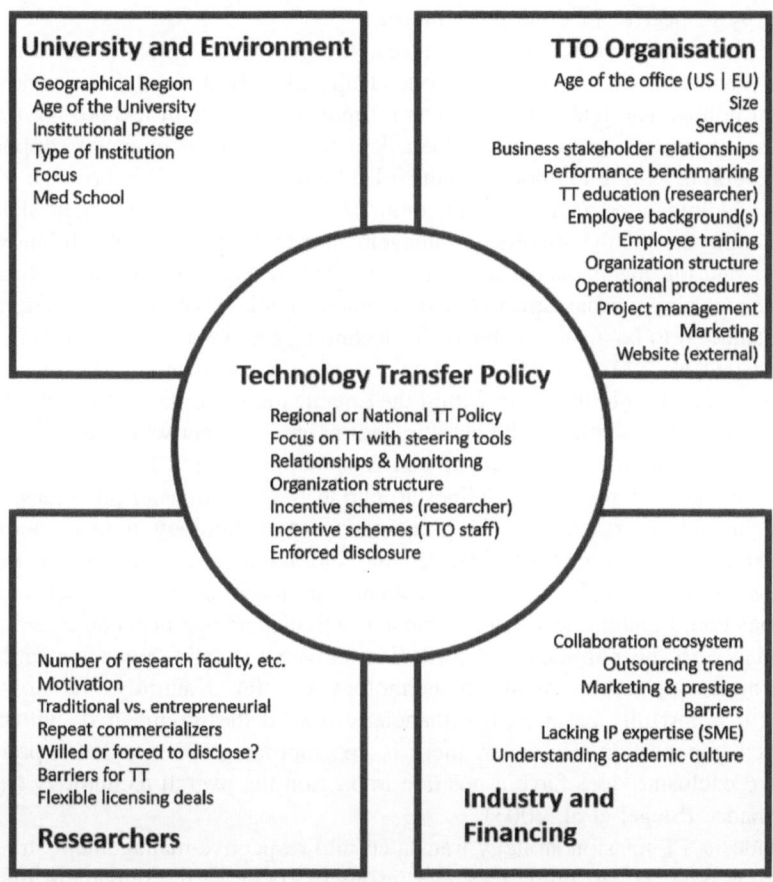

Fig. 1 The five dimensions and influence factors of technology transfer

them is essential for the prosperity of a technology transfer office and therewith presents a target for intervention opportunities that can be addressed by a multitude of different capacity building strategies for TTOs.

2 Technology Transfer Policy

Technology transfer as the 3rd mission of universities besides teaching and research has become a widely adopted policy framework worldwide. In this realm, technology transfer should be recognized as a contribution to the excellence in research and teaching, sustainably backed by universities' management. Such an "entrepreneurial university" is recognized as an important building block of an entrepreneurial society (Audretsch 2014). Empirical evidence shows that TTOs that receive strong institutional support are more successful (Debackere and Veugelers 2005; O'Shea et al. 2005). This section looks into how the success of a TTO can be shaped by respective institutional policies.

It is beneficial for the TTO, if the research organization officially declares technology transfer as its strategic mission, ideally directly derived from regional or national policy. The real value a research organization sees in technology transfer processes, however, is not only reflected in the amount of financial support the organization is willing to allocate to the TTO budget but as well in how the TTO is managerial integrated in the organization. If the strategic endorsement shall not remain a lip service, the university management actively has to establish innovation focused steering mechanisms, to which the TTO management has to report. Sustainable reporting mechanisms to the top management level of the organization have been identified to be a success factor for technology transfer (York and Ahn 2012; Bercovitz et al. 2001), preferably such reporting mechanisms allow the TTO a significant amount of autonomy within the organization in order to sustain a certain level of freedom essential to the entrepreneurial environment within the TTO.

The organization's management can also support the TTO by establishing institutional procedures and guidelines around technology transfer processes, which should incentivize researchers in participating in technology transfer activities. Establishing the 3rd mission requires the enforcement of technology transfer activities without weakening the excellence in teaching and research. Tenure decisions could include technology transfer activities as one decision criteria, in a fair balance to the publication record. The university could sponsor awards and prizes honouring achievements in technology transfer. Naturally, the university could (or is lawfully required to) financially reward the produced IP with a significant compensation. However, there is no empirical evidence that a policy of enforced enclosure does have a positive impact on the overall technology transfer performance (Siegel et al. 2003).

Besides a TT mission strategy translated into respective management structures as well as policies and guidelines, the resources spent by the university management for the TTO are another obvious indication for how much technology transfer is valued within the organization. Having a critical mass of permanent, full-time employees dedicated to technology transfer activities is an important indicator that may be directly derived from a TT oriented university management policy. Additionally, the university management should invest into these human resources (HR), by applying appropriated HR tools, such as performance based salary components and dedicated career development plans.

In summary, a coherent strategy towards the 3rd mission on the national, the regional, and the university level is necessary for successful technology transfer. The university management has several options for proving its commitment towards the 3rd mission. Without clear and direct support from the university management, the TTO at an institution will hardly be able to succeed.

3 University & Environment

The primary role of a TTO is to manage and perform technology transfer activities (AUTM 2013). In the technology transfer process, the single TTO constitutes the essential junction for bringing the different partners (public research institution, researcher, industry, investors, and other intermediaries) together—often literally, in in-person meetings organized by TTO staff and hosted at the TTO. This section explores how the location of the TTO in a geographical region, a local network, or associated with a specific university affects its effectiveness to do so.

Universities are part of a system of innovation and therewith influenced by various symbiotic relationships (Freeman 1995). Innovation can be promoted by national science politics, e.g. through solid IP laws and tax incentives for start-ups as well as governmental research funding for technology transfer activities. All of these directly affect the commercialization success of a TTO. For a more detailed discussion of how national and institutional political initiatives can support technology transfer offices in their work, see Sect. 2.

On the regional level, the innovation generating interactions between university, industry, and government have been described as a "triple helix" system with the overall function to generate, diffuse, and use knowledge (Etzkowitz and Leydesdorff 1995). This aim is realized through relationship building as well as the creation of knowledge, innovation, and consensus spaces, wherein the individual innovators (researcher, business representative, policy maker, student, entrepreneur, business angel, venture capitalist) have to consolidate inter-institutional processes (Ranga and Etzkowitz 2013). Accordingly, the TTO should serve as the central hub for these regional development relationships at the intersection of the university, industry, and government spheres. The "triple helix" model appears to be especially applicable to Europe, where small and medium enterprises (SMEs) are predominant, and an increasing number of SMEs recognize the importance of vertical and horizontal networks for collective learning and innovation within their operative region (Cooke 2002). Public research organizations in a specific region generate intellectual property adding to the intangible "proximity capital" of this region (Crevoisier

1997). There is "knowledge spillover" from the PROs (Hellerstedt et al. 2014); for example, highly educated graduates join the regional work force or start their own company creating further employment opportunities in the region as well as taxable revenue. With former students now working at one of the local companies, social ties between industry employees, academic entrepreneurs, and university researchers are growing (Heblich and Slavtchev 2014). Over time, these personal relationships evolve naturally, until eventually the initial knowledge transfer from the university into industry may be reflected back to the academic institution in the form of e.g. sponsored research projects, potential license agreements, or student training internships. Through these symbiotic processes the university city may evolve into an entrepreneurial hub of regional excellence further increasing the attractiveness of the university and the region, drawing talent (national, international) and venture capital (VC) firms, and therewith fuelling the transformation of the region into a world class entrepreneurial ecosystem (Ranga and Etzkowitz 2013), wherein the TTO acts as the liaison between the individual innovators. Respectively, technology transfer may be used as a key vehicle to promote the development of a specific region (Shaw and Allison 1999). There are various examples in the US and Europe, where an entrepreneurial university acts as a catalyst for regional economic and social development: e.g. Boston Area (Route 128, MA, US), Silicon Valley (CA, US), Cambridge Science Park (Cambridge, UK), Kista City (Stockholm, Sweden), Innovation & Incubation Center (KU Leuven, Belgium), or Catalonia Innovation Triangle (Spain; Urbano and Guerrero 2013). As part of a structured innovation organization initiative, affordable physical infrastructure may be provided, together with mentoring programs and speaker series to build the entrepreneurial skill set, or voucher services (legal, finance, accounting, marketing, management) to support young start-ups. Most of these services may be offered or coordinated by the TTO located in the respective region. The TTO may benefit from this large amount of interaction, as it may result in more transactions, ultimately leading to more deals, more spin-offs, and higher licensing revenue. A few voices, however, warn that the creation of an innovation system around a university may be associated with various constraints such as lack of diversity and therewith robustness (Sandström et al. 2016), while in light of the increasing globalization the importance of such regional networks may diminish in the future, when knowledge may be created and shared freely through ties of global network allies (Whittington et al. 2009) and the expertise of the single researcher may become the dominant driving factor for commercialization activities (Cummings and Teng 2003).

On the level of the single university, the success of its TTO strongly depends on the history of the academic institution, its research orientation with associated commercialization potential, in addition to its faculty quality and size as well as the number of graduate students and postdocs (O'Shea et al. 2005). While the majority of licensing revenue is often based on 1 in 1000 invention disclosures, that one big success likely originates in the applied, computer, or life sciences (AUTM 2013). Approximately 90% of companies conducting life science research in the US report to have relationships with academic institutions, of which 95% directly support research activities (York and Ahn 2012), with similar numbers in the engineering and

IT sector. Consequently, a TTO at a university where research is conducted in these areas, potentially in association with a medical school and affiliated hospital(s), has access to a larger pool of research products with high commercialization potential (Blumenthal et al. 1996). Furthermore, older universities appear to have clear advantages in terms of stakeholder relationships and acceptance, because the curation of a large alumni network and trustworthy relationships with the local and broader business community as well as establishing credibility with VCs and other potential funding partners, such as foundations or network agencies, takes time. In addition, these institutions have had more time to focus on technology transfer activities, may have established their TTO earlier, may have implemented institutional policies to support knowledge mobilization activities a long time ago, and in the process may have attracted, educated, and retained experienced TTO personnel over the years.

With the age of an institution, its global visibility often increases as well, as the university had more time to build their brand of academic excellence, e.g. reflected in high scores in national/global academic rankings. The so-called "halo effect" (Sine et al. 2003) radiates from such an institution promising trustworthiness to industry collaborators and investors which leads to an increased number of collaborative research projects and licensing agreements as well as funded spin-offs so that the performance of the respective TTO shines brightly under the institution's halo, too. A university's prestige is as well likely to attract "star scientists" who increase the research product (Zucker et al. 1998), leading to more high quality invention disclosures to the TTO and therewith a growing chance for a high success technology. While the majority of US universities have only modest technology transfer related success and income, a relative small number of highly advanced and established institutions that are located in extraordinary regional ecosystems and foster a strong entrepreneurial campus culture create the bulk of commercialization success (AUTM 2013) with negligible differences between public and private universities (Anderson et al. 2007).

In conclusion, a TTO at a prestigious and entrepreneurial university employing renowned faculty in the applied, computer, and life sciences that is located in an economically strong and geographically attractive region has certain advantages in regard to success parameters. As the following section will show, however, an innovative TTO with highly motivated and capable personnel may be able to overcome some of the perceived historic barriers and create new entrepreneurial hubs, especially if supported by an innovation promoting university culture and national policy.

4 Organization of the Technology Transfer Office

As discussed in the previous section, the TTO is playing an essential role as a hub in the regional innovation ecosystems. Obviously, the organization structure and procedures within the TT offices are important influence factors for the success of their technology transfer activities and will be explored in this Section.

Comparable to the positive impact of the universities' age, older TTOs may have advantages in regard to stakeholder relationships easing their technology transfer activities; however, there is no empirical evidence that age is a significant success factor (York and Ahn 2012), at least in the US. This favors the situation in Europe. According to the EKTIS study (2010–2012) TTOs on European universities have an average age of 14 years, only 21% of them have been established before 1990 and, even more importantly, 52% have been established after 2000 only (Arundel et al. 2013). Partially, these numbers can be explained by the situation of the transformative economies in Middle and Eastern Europe, but even in Western and Northern Europe the majority of TTOs do not have a long historical record. Independent from the age of the TTOs, the maturity and strength of the relationships the single office has built with the local and broader business community as well as with venture capitalists and other potential funding groups are strongly beneficial for the technology transfer success (York and Ahn 2012; Shane 2002). These relationships, preferably reaching out beyond the local community, are strong predictors of the likelihood for future deals.

Technology transfer is a resource intensive people business, where trustful and long-term personal relationships have to be built through face-to-face interactions. In order to be successful, TTOs need a critical mass. Empirical evidence shows, that in the US, TTOs with more than 15 full-time employees are significantly more successful than their smaller counterparts (York and Ahn 2012). While in Europe, 15 full-time employees on average are employed at a TTO, about 42% of the TTOs in Europe have only 5 employees and 17% have only 2 staff members (Arundel et al. 2013). These smaller TTOs are far from having a critical mass for successful operations.

Besides the small staff size, the usual high staff turnover rates in TTOs are counterproductive for the long-term relationships into the regional ecosystem and for securing the knowledge capital built by the staff. A low turnover, stable job perspectives for the staff, as well as career development plans are needed to retain well-trained personnel and to secure the success potential of the TTO. Following this argument, employee training for TTO staff, including the achievement of professional recognition through certification programs (as done by AUTM, ASTP-PROTON, or LES) are important to increase employee loyalty and reduce fluctuation.

Regarding the formal qualification of the personnel, a mixture of backgrounds in technology, management and business administration as well as law and finance is considered ideal (Arundel et al. 2013). Technology transfer professionals require a strong scientific background and a solid understanding of patent and contract law in addition to marketing and negotiation skills. In part, technology transfer is a marketing and sales exercise, hence skilled negotiators and marketers, being able to seek paying customers for inventions, are required. This skill set is usually underrepresented in TTOs, mostly because TTOs—due to their organizational culture and lack of financial incentives—are not the preferred environment for skilled and successful sales personnel, which makes honing these skills for the TT professionals even more important.

As the comprehensive taxonomy of technology transfer services, which will be introduced in the remainder of the book, shows, TTOs have a variety of different services in which they can specialize. Empirical evidence demonstrated that TTOs, which are focusing on licensing and spin-off generation, are performing better than TTOs that do have a wider focus of activities (Conti and Gaule 2008).

As depicted in Fig. 2, intermediaries do have several options for business models to position themselves with the research organizations they serve (Bader et al. 2011). While the close integration of the TTO into the university (quadrant I) is the predominant model in the market, all other relationship models can be observed in the market as well. The current establishment of the SATT structures in France is a good example of the business model in quadrant IV, where one, legally independent, TTO is serving a group of universities (Esteve et al. 2013). Empirical evidence shows that one large TTO serving (ideally) one campus is on average the most successful model, while multi-campus systems with separate offices at each campus seem to lead to organizational and communicational difficulties (York and Ahn 2012; Gulbranson and Audretsch 2008; Lambert 2003).

Besides their position against the universities they are serving, TTOs do also have to decide which of the services they are offering are realized in-house and which of these can be provided by sub-contracted third parties or shared with other TTOs in a cluster of expertise. Literature indicates that shared services could preferably be licensing negotiations, market research for new technologies, IP marketing, IP management or spin-out creation, while services in-house should preferably be raising academic awareness on technology transfer, negotiation of collaborative research contracts, reach-out to businesses, and consultancy contracts (Lambert 2003). Regardless which concrete model of service provision a TTO decides on, those with a strategic approach to operations seems to be more successful (Lambert 2003).

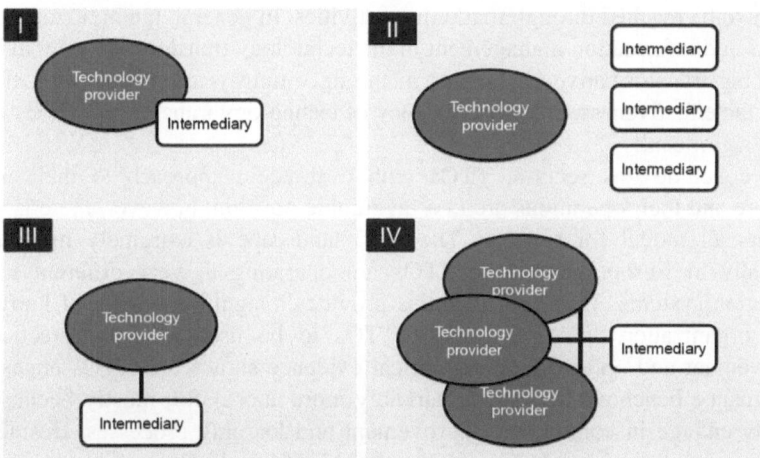

Fig. 2 Schemes for business models between technology providers and intermediaries. *Source* Bader et al. (2011)

Transparent operations with standardized interfaces to all stakeholders and proven service levels appear to benefit the performance of TTOs. This is strongly interlinked with the project management patterns the TTOs employ to serve the customers. Literature differentiates between the pure "Marathon", the planned "Hybrid" and the "Medley" models (Bercovitz et al. 2001). Similar to a key account, the Marathon model provides a single point of contact during the full customer interaction, while the planned hybrid hands over the customer interaction between different teams in the TTO at defined stages in the process. In the Medley model, interdisciplinary teams handle various topics with the customer in parallel. Pure Marathon or planned Hybrid models for managing projects have empirically proven to be more successful than Medley models (Bercovitz et al. 2001).

As mentioned earlier, in most TTOs marketing is one of the underdeveloped fields of operations. In general, TTOs do have three types of assets they can promote in marketing activities: (1) the technology base on which they can access, (2) the body of successful licensing deals and spinout creations in the past, and (3) the quality and excellence of their practices and procedures. Not surprisingly, TTOs with a strategic approach towards marketing are more successful (Arundel et al. 2013). TTOs do use a multitude of marketing channels like workshops and conferences, personal contacts of TTO staff, press statements, newsletters, trade fairs, direct mailing, printed magazines, external technology intermediaries and consultants, and most prominently, their websites (Arundel et al. 2013). Especially the value of the TTO website should not be underestimated, as it is an important differentiator between TTOs in establishing virtual proximity with potential collaborators and stakeholders. TTOs with easy-to-use, comprehensive, informative websites that appeal to internal and external stakeholders tend to be more successful (York and Ahn 2012). With the realm of business platforms, such as LinkedIn, or communication channels, such as Twitter, the marketing activities of the TTO should be integrated into the digital experiences and digital practices of all stakeholders to be reached through marketing activities. In general, the digitization of the process and information management in the technology transfer practice will be one of the big transformative challenges in the upcoming years. As a result, the outreach, the effectiveness, and the efficiency of technology transfer operations will be scaled significantly.

To conclude this section, TTOs with a strategic approach to their internal structure and their operations are in a favourable position; however, there is no one size fits all model for success. The TTO landscape is extremely multifaceted, especially in Europe where the TTOs are operating in very different national innovation systems. This vast expertise provides a significant body of knowledge about organization and operation of TTOs to be used as good practices for improvement and optimization. Empirical evidence shows that TTOs engaging in performance benchmarking are remarkably more successful, mostly because they actively engage in continuous improvement and learning processes. Besides performance benchmarking (such as e.g. the AUTM technology transfer rankings) using and adopting best practices is a very valuable learning tool for TTOs.

5   Researcher Involvement

Before technology can be transferred, it has to be created first. Therefore, the researcher as the technology supplier and developer is critical for the technology transfer process and the success of a TTO. At a university, researchers are staff and faculty members as well as the students and postdocs they are training. How the pure quantity of researchers at a university leads to more inventions and therewith to more technologies with commercial potential has been discussed in Sect. 3, while in this Section the soft factors that motivate researchers to participate in technology transfer processes are addressed.

The universal idea of scientific research as a serious and methodologically planned quest for truth was first expressed by Wilhelm von Humboldt and later studied by Robert Merton, who summarized the concept in four terms: (1) Communalism: free access to knowledge, (2) Universalism: blind peer-review for quality assurance, (3) Disinterestedness: in personal gain, such as financial benefit, and (4) Organized Scepticism: continuous questioning of the scientific methodologies and institutions (Anderson et al. 2010). Over centuries these values have shaped society's view of the role of researchers as well as the function of public research institutions and laid the foundation for the mutual contract that public research funds should be used to sponsor research for the advancement of society—although nowadays voices calling for a new social contact become louder (Martin 2012; Slaughter 2001).

Publication of research results, as expressed in the first Mertonian concept, remains the key success factor to a distinctive career in science, where achievement is measured in publication volume (presentations, journal article, books) further applying various metrics, such as e.g. journal impact factors or the h-index/core/median. These are some of the wide used tools that allow ranking of the individual researcher to determine future funding support and performance towards promotion along the academic career path. Hence, publishing the research work is essential for advancement, especially for early-stage researchers and junior faculty members, which in addition to mandatory departmental service leaves not much time for commercialization activities. Not surprising, full professors who have been tenured for more than 10 years are more likely to participate in TT activities (Link et al. 2007). The impact the researcher's career stage has on the TT process can only be overcome by adapting the university's promotion and tenure remuneration systems to actually value commercialization activities in addition to the traditional publication metrics (Siegel and Wright 2015; Stevens et al. 2011), which may be the first step towards an entrepreneurial campus culture (see as well Sect. 2).

Furthermore, the peer-review process and the critical discussion of scientific ideas, as expressed in the second and fourth Mertonian concept, are key elements of a research career. Driven by the urge to solve a puzzling question, the researcher collects data and eventually publishes the findings, therewith gaining recognition in the research community (awards, lecture invitations, prestigious membership(s) in scientific society), which increases the likeliness for funding to pay for future research endeavours. Accordingly, the researcher lives a life that values the "puzzle" (intrinsic satisfaction), the "ribbon" (extrinsic reputational/career reward), and the "gold" (extrinsic financial reward) (Lam 2011); however, which of the three may be the dominating factor for a researcher to participate in TT actives depends on the individual.

Similarly, how exactly a researcher may interpret the third Mertonian principle is subject to the individual's personal beliefs. Some researchers adhere strictly to the traditional idea of basic research and follow their belief that the search for scientific truth is incompatible with personal financial profit (Louis et al. 1989). For these pure traditionalists, technology transfer activities pose an inherent conflict of interest, and when forced to commercialize, they may feel as victims of the cutthroat capitalism (Slaughter 2001). On the other side of the spectrum stands the personality of the entrepreneurial researcher, who interprets the social contract differently and is motivated by giving back to society more directly, e.g. by bringing a new drug to the market, or starting a company to create employment opportunities for others. This type of researcher is intrigued by research questions that aim to solve real-life problems, which often also possess considerable commercial value, something that the entrepreneurial researcher is not conflicted over (Lam 2011). A sub-class are "repeat commercializers", who according to a study by Hoye and Pries (2009) represent only about 12% of the faculty at an institution but account for about 80% of its commercialized innovations, which is further supported by the principle of performance persistence in entrepreneurship (Gompers et al. 2008). The majority of researchers though fall between the two extremes of the traditional researcher and the entrepreneurial researcher and consider themselves more of a pragmatic traditionalist, who favours the distinction between academe and industry but also recognizes the necessity for collaboration between the two, or a hybrid, who thrives in academe-industry collaborations but understands the need to maintain boundaries (Lam 2011). Due to ongoing changes of the public funding landscape accompanied by devastating budget cuts, many researchers begin to consider TT out of necessity to subsidize their research funding and e.g. higher royalty shares for researchers have been associated with greater TTO licensing income (Link and Siegel 2005).

Finally, if and how motivated a researcher may be to disclose an invention to the TTO depends on the individual's perception of the value of IP, the ease of the TT process, and the dominating opinion about TT on campus (Owen-Smith and Powell 2001), as especially individuals with high group identification base their commercialization intentions on the function of social norms and need to fit in with their workplace peers (Obschonka et al. 2012). Moreover, differences in the researchers' motivation exist between science-based and development-based regimes (Gilsing et al. 2011), while physical scientists may patent for freedom of action, life scientists may search IP protection to gain a strategic advantage in negotiations with industry (Owen-Smith and Powell 2001). A TTO with a strong presence on campus and long-time staff members, who have built relationships with key members of the university research community, may have created the perception of the TT process as a pleasant experience and may be openly received by the individual researcher. The TTO plays a key role in building a campus culture that values commercialization activities; for example, by providing IP and entrepreneurship training to their campus researchers and negotiating deals beneficial to the university as well as to the individual researcher. Wherein fostering a researcher's entrepreneurial mindset by providing training and networking experiences showed to be a stronger determinant than expected economic or reputational gain in motivating researchers to start companies (Goethner et al. 2012).

In summary, a TTO at a large academic institution with a strong focus on research is supplied with greater amounts of university research product, increasing the chances of a commercial success technology. Moreover, if an academic institution is actively promoting a campus culture that supports academic as well as commercial endeavors, researchers at this institution will be motivated to participate in technology transfer activities and commercialization endeavors for various individual economic and psychological reasons, especially if they have established a good personal relationship with their TTO.

6 Industry & Financing

For technology transfer capacity building the TT ecosystem plays an important role. This ecosystem involves university, industry, and government and has been described as a "triple helix" (Etzkowitz and Leydesdorff 1995). In this section the interaction between university and financial stakeholders is examined.

According to previous research by O'Shea et al. (2005) the TT success depends on a successful engagement with industry and financial stakeholders to attract funding for research and development (R&D) and commercialization processes. According to the EKTIS study (2010–2012), large as well as SMEs engage with public research organizations (Arundel et al. 2013). Nevertheless, the majority of university licenses (55%) are issued to large firms.[1] Interestingly, proximity is apparently not an issue while engaging with large enterprises. As innovation is going global, large enterprises built their R&D centres in the most important markets, a choice that is not affected by their national heritage.

Nonetheless, proximity is important when engaging with SMEs, where local and regional agencies develop science-based clusters (Lambert 2003). The reason behind the proximity factor is information barriers. In general, innovation business communities are well informed and bigger communities have more resources for getting engaged with PROs out of home countries or regions. Smaller enterprises have resource restraints, and therefore information barriers. As a result, the EKTIS (2010–2012) dataset showed proximity importance only for SMEs and PRO engagements. In the industry university ecosystem for most of the licensing deals made (55%) research partner proximity is not important (Arundel et al. 2013).

---

[1]Large firm: more than 250 employees.

Increasing global competition and technological change have forced companies to move away from exclusive in-house R&D. Instead, we can witness a trend among industry to focus on its core strengths and to outsource a growing proportion of their activities (Lambert 2003). One of the drivers for that shift can be identified as the organizational changes in technology companies. The way technology producers function has changed since the 1980s. Three decades ago, industry firms functioned as a collection of divisions based in various regions, business units, or product lines. Nowadays, technology producers are organized often as an array of specialized business-units (e.g., procurement, manufacturing, research and development, marketing, sales, IP management, and distribution, etc.). In this setting, they have bit by bit opened up their innovation processes resulting in technology producers outsourcing some of the fragments of the innovation processes (Palmisano 2006). While the fact that technical expertise is often leaving the technology producers quickly has been identified as one of the major drawbacks of the outsourcing trend, the new setting of industry opening up to collaborations with university research labs offers many advances, such as access to brain power and excellence across a wider range of disciplines needed to answer the increasingly more complex research questions of today, access to specialized consultancy and public funding as means to leverage research funds, as well as a chance to spot young talent and recruit first, and an overall reduced risk to expand pre-competitive research (Lambert 2003; Arundel et al. 2013). Companies involved in collaborations with universities are more likely to broaden their range of goods/services, open new markets or increase their market share, in addition the collaboration with top university researchers may create scientific prestige to fuel marketing efforts (Zucker et al. 1998).

Although industry-university collaborations have many benefits for both sides, many barriers appear to exist on the intercultural and interpersonal level as summarized in Table 1.

Table 1 Perceived barriers of joint research projects between university and industry

Barriers identified by industrial researchers
• Joint research projects with universities or PROs are difficult to manage and/or involve high overhead costs
• The results of joint research projects with universities or PROs imply a significant risk of leaks to competitors
• Knowledge developed in universities and PROs is too general to address our specific knowledge needs
• Being involved in the application of knowledge developed in universities or PROs is too costly (either in terms of time or money)
• Relevant knowledge developed in universities and PROs is difficult to locate (e.g., finding the right publications or people)
• Researchers working in universities or PROs do not fit in well with our corporate culture
• Knowledge developed in universities and PROs is too theoretic to be useful in our particular case

Source Gilsing et al. (2011)

In the innovation ecosystem, intellectual property and related policies play an important role (Tonisson and Maicher 2012; Tonisson et al. 2016) and occasionally are considered an important barrier to university-industry engagement. The company interviews made in the EKTIS study (Arundel et al. 2013) as well as other literature (Perkmann and West 2014; Siegel et al. 2003), highlight the high transition costs of research partnerships caused by the even more ambitious IP policies in Europe. Such policies, designed to rise the commercialization costs of research, discourage industry to engage with universities and thus create barriers for academics in their attempts to work with industry (Bruneel et al. 2010; Hertzfeld et al. 2006). Bureaucratic application processes and lack of clarity over ownership of IP in research collaborations appear to be often a source of frustration for TTOs (Lambert 2003). Outsourcing IPR related tasks to IPR service providers may help to ease this situation (Tonisson et al. 2016).

In conclusion, both, industry and PROs, could benefit even more from collaboration activities, if universities understood the corporate culture better and industry understood the university R&D culture better. Intercultural communication appears to play an important role in overcoming the perceived barriers.

7 Conclusion

The success of a TTO is determined by the performance evaluation criteria chosen. While many TTOs strive to become profit centers for their research organization, which often might be the main motivator for the respective research organization to fund a TTO in the first place, the volatile nature of the licensing market makes it challenging to project profits or losses. Measures of reward for a pure revenue model TTO include: gross revenue, net revenue, equity cash flow, and new industry subsidized research partnership funds (Batalia 2006). There is proof that the revenue model works, as regression analysis of data from AUTM annual reports confirmed that TTOs that make generating royalty income their top priority produce better financial results than TTOs that do not (Friedman and Siberman 2003). Nevertheless, focusing solely on financial results is one-sided and prone to neglect other significant performance factors (Axanova 2012).

The prosperity of a technology transfer office is built on five fundamental pillars: policy, university & environment, TTO internal organization, researcher involvement, and industry & financing (Fig. 1). Each one of these pillars is essential to the overall success of a single office and therewith provides an important intervention opportunity for increasing the capacity building of a TTO. The different stakeholders (public/government, research institutions, TTO directors, researchers, industry, investors) may choose to intervene in one of their areas of expertise to work together with the TT professionals in the respective TTO towards increasing the achievements of said office.

This chapter meant to provide a starting point for anyone who is planning to set-up a new TT program or to improve the performance of an existing one. The carved out five fundamental pillars and the identified influence factors of performance success for each of them may be targeted by a multitude of different capacity building strategies. The next chapters of this book will show how Progress-TT narrowed down its own approach.

References

Anderson MS (2001) The complex relations between the academy and industry: views from literature. J Higher Educ 72:226–246

Anderson MS, Ronning EA, De Vries R, Martinson BC (2010) Extending the Mertonian norms: scientists' subscription to norms of research. J Higher Educ 81:366–393

Anderson TR, Daim TU, Lavoie FF (2007) Measuring the efficiency of university technology transfer. Technovation 27:306–318

AUTM (2013) AUTM U.S. licensing activity survey: FY2012 highlights. Online (behind paywall). http://www.autm.net/resources-surveys/research-reports-databases/licensing-surveys/archived-licensing-surveys/

Arundel A, Es-Sadki N, Barjak F, Perrett P, Samuel O, Lilischkis S (2013) Knowledge transfer study 2010–2012: final report. Publication Office of the European Union. https://ec.europa.eu/research/innovation-union/pdf/knowledge_transfer_2010-2012_report.pdf

Audretsch DB (2014) From the entrepreneurial university to the university for the entrepreneurial society. J Technol Transf 39:313–321

Axanova L (2012) U.S. academic technology transfer models: traditional, experimental and hypothetical. Les Nouvelles 2:125–137

Bader M, Gassmann O, Jha P, Ziegler F, Maicher L, Posselt T, Preissler S, Ruether F, Wabra S (2011) Creating a financial market for IPR: final report. https://www.imw.fraunhofer.de/content/dam/moez/de/documents/Studien/study_ipr-in-europe.pdf

Batalia M (2006) Avant-garde technology transfer at a midsize, private university. In: AUTM technology transfer practice manual, vol 2, Part 1, Chapter 2.2c

Bercovitz J, Feldman M, Feller I, Burton R (2001) Organizational structure as a determinant of academic patent and licensing behavior: an exploratory study of Duke, Johns Hopkins, and Pennsylvania State Universities. J Technol Transf 26:21–35

Blumenthal D, Causino N, Campbell E, Louis KS (1996) Relationships between academic institutions and industry in the life sciences—an industry survey. N Engl J Med 334:368–373

Bruneel J, D'Este P, Salter 'A (2010) Investigating the factors that diminish the barriers to university-industry collaboration. Res Policy 39:858–868

Conti A, Gaule P (2008) The CEMI survey of university technology transfer offices in Europe report. CEMI report. http://cemi.epfl.ch/page-30722-en.html

Cooke P (2002) Regional innovation systems: general findings and some new evidence from biotechnology clusters. J Technol Transf 27:133–145

Crevoisier O (1997) Financing regional endogenous development: the role of proximity capital in the age of globalization. Eur Plan Stud 5:407–415

Cummings JL, Teng BS (2003) Transferring R&D knowledge: the key factors affecting knowledge transfer success. J Eng Technol Manage 20:39–68

Debackere K, Veugelers R (2005) The role of academic technology transfer organizations in improving industry science links. Res Policy 34:321–342

Esteve C, Carboni N, Laminae V (2013) SATT—The new French model for TTO. A presentation for the TTO circle. https://ec.europa.eu/assets/jrc/events/20130627-tto-circle/jrc-20130627-tto-circle-carboni.pdf

Etzkowitz H, Leydesdorff L (1995) The triple helix—university-industry-government relations: a laboratory for knowledge based economic development. EASST Rev 14:14–19

Etzkowitz H, Webster A, Healey P (eds) (1998) Capitalizing knowledge: new intersections of industry and academia. State University of New York Press, Albany, NY

Freeman C (1995) The national system of innovation in historical perspective. Camb J Econ 19: 5–24

Friedman J, Siberman J (2003) University technology transfer: do incentives, management, and location matter? J Technol Transf 28:17–30

Gilsing V, Bekkers R, Bodas Freitas IM, van der Steen M (2011) Differences in technology transfer between science-based and development-based industries: transfer mechanisms and barriers. Technovation 31:638–647

Goethner M, Obschonka M, Silbereisen RK, Cantner U (2012) Scientists' transition to academic entrepreneurship: economic and psychological determinats. J Econ Psychol 33:628–641

Gompers PA, Kovner A, Lerner J, Scharfstein DS (2008) Performance persistence in entrepreneurship. Harvard Business School Working Paper 09-028

Gulbranson CA, Audretsch DB (2008) Proof of concept centers: accelerating the commercialization of university innovation. J Technol Transf 33:249–258

Heblich S, Slavtchev V (2014) Parent universities and the location of academic startups. Small Bus Econ 42:1–15

Hellerstedt K, Weinberg K, Frederiksen L (2014) University knowledge spillovers and regional start-up rates: supply and demand-side factors. In: Corbett A, Siegel DS, Katz JA (eds), Academic entrepreneurship: creating an entrepreneurial ecosystem, Volume 16: Advances in entrepreneurship, firm emergence and growth. Emerald Group Publishing Ltd., Bingley, pp 137–168

Hertzfeld HR, Link AN, Vonortas NS (2006) Intellectual property protection mechanisms in research partnerships. Res Policy 35:825–838

Hoye K, Pries F (2009) 'Repeat commercializers', the 'habitual entrepreneurs' of university-industry technology transfer. Technovation 29:682–689

Lam A (2011) What motivates academic scientists to engage in research commercialization: 'gold', 'ribbon' or 'puzzle'? Res Policy 40:1354–1368

Lambert R (2003) Lambert review of business-university collaboration: final report. http://www.eua.be/eua/jsp/en/upload/lambert_review_final_450.1151581102387.pdf

Link AN, Siegel DS (2005) Generating science-based growth: an econometric analysis of the impact of organizational incentives on university-industry technology transfer. Eur J Fin 11:169–181

Link AN, Siegel DS, Bozeman B (2007) An empirical analysis of the propensity of academics to engage in informal technology transfer. Ind Corp Change 16:641–655

Louis KS, Blumenthal D, Gluck ME, Stoto MA (1989) Entrepreneurs in academe: an exploration of behaviours among life scientists. Admin Sci Q 34:110–131

Martin BR (2012) Are universities and university researchers under threat? Towards an evolutionary model of university speciation. Cambridge J Econ 36:543–565

Obschonka M, Goethner M, Silbereisen RK, Cantner U (2012) Social identity and the transition to entrepreneurship: the role of group identification with workplace peers. J Vocat Behav 80:137–147

O'Shea RP, Allen TJ, Chevalier A, Roche F (2005) Entrepreneurial orientation, technology transfer and spinoff performance of U.S. universities. Res Policy 34:994–1009

Owen-Smith J, Powell WW (2001) To patent or not: faculty decisions and institutional success at technology transfer. J Technol Transf 26:99–114

Palmisano SJ (2006) The globally integrated enterprise. Foreign Affairs Web 27 July 2017. https://www.foreignaffairs.com/articles/2006-05-01/globally-integrated-enterprise

Perkmann M, West J (2014) Open science and open innovation: sourcing knowledge from universities. In: Link AN, Siegel DS, Wright M (eds), The Chicago handbook of university technology transfer and academic entrepreneurship. University of Chicago Press, Chicago, pp 41–74

Powers JB (2003) Commercializing academic research: resource effects on performance of university technology transfer. J Higher Educ 74:26–50

Ranga M, Etzkowitz H (2013) Triple helix systems: an analytical framework for innovation policy and practice in the knowledge society. Ind Higher Educ 27:237–262

Sandström C, Weinberg K, Wallin MW, Zherlygina Y (2016) Public policy for academic entrepreneurship initiatives: a review and critical discussion. J Techno Transf. https://doi.org/10.1007/s10961-016-9536-x

Segatto-Mendes AP, Mendes N (2006) University-Industry technological cooperation for energy efficiency: a case study. Braz Admin Rev 3:31–45

Shane S (2002) Executive forum: university technology transfer to entrepreneurial companies. J Bus Venturing 17:537–552

Shaw JK, Allison J (1999) The intersection of the learning region and local and regional economic development: analysing the role of higher education. Reg Stud 33:896–902

Siegel DS, Waldman D, Link A (2003) Assessing the impact of organizational practices on the relative productivity of university technology transfer offices: an exploratory study. Res Policy 32:27–48

Siegel DS, Wright M (2015) Academic entrepreneurship: time for a rethink? Br J Manage 26:582–595

Sine WD, Shane S, DiGregorio D (2003) The halo effect and technology licensing: the influence of institutional prestige on the licensing of university inventions. Manage Sci 49:478–496

Slaughter S (2001) Professional values and the allure of the market. Academe (Sep–Oct) 1

Stevens AJ, Johnson GA, Sandberg PR (2011) The role of patents and commercialization in the tenure and promotion process. Technol Innov 13:241–248

Tonisson L, Maicher L (2012) Patents, their importance and valuation methods. Fraunhofer Working Paper 3/2012. http://publica.fraunhofer.de/eprints/urn_nbn_de_0011-n-2264535.pdf

Tonisson L, Millien R, Maicher L (2016) Shortcomings on the market for intellectual property. Fraunhofer Working Paper 1/2016. https://www.imw.fraunhofer.de/content/dam/moez/de/documents/Working_Paper/Working-Paper_Shortcomings%20on%20the%20market%20for%20intellectual%20property.pdf

Urbano D, Guerrero M (2013) Entrepreneurial universities: socioeconomic impacts of academic entrepreneurship in a European region. Econ Dev Q 27:40–55

Wayne KT (2010) Determinants of commercial innovation for university technology transfer. J Behav Stud Bus 2:1

Whittington KB, Owen-Smith J, Powell WW (2009) Networks, propinquity, and innovation in knowledge-intensive industries. Admin Sci Q 54:90–122

York AS, Ahn MJ (2012) University technology transfer office success factors: a comparative case study. Int J Technol Transf Com 11:26–50

Zucker LG, Darby MR, Armstrong J (1998) Geographically localized knowledge: spillovers or markets. Econ Inq 36:65–86

The Dynamics of Inter-organizational Hybrid Partnerships in Technology Transfer

Fabiola Bertolotti, Elisa Mattarelli and Paula Ungureanu

Abstract

Drawing on the literature on inter-organizational and hybrid partnerships, we put forth a process-based perspective on the evolution of regional innovation systems (RIS), with particular attention to the changing role of TTOs throughout the RIS lifecycle. We theorize on how perceptions of environmental turbulence (volatility, uncertainty, complexity and ambiguity, in short VUCA) may influence partners' decisions to adopt a given organization model for the broker/TTO that manages the partnership. We show that perceptions of environmental turbulence may lead to a set of possible decision pathologies at the partnership level that interfere with the organizational structure of the TTO. We suggest that perceptions of turbulence and decision pathologies play an important part in explaining RIS may deviate from the intended direction or produce outcomes that are unexpected.

1 Introduction

In the last decades, we have witnessed an increasing attention of scholars and policy makers alike toward a better understanding of the processes through which intellectual property can be moved more effectively into industry through the creation of Technology Transfer Offices (TTO) and how existing TTOs can improve their performance and that of the regions in which they are immersed.

On a parallel basis, the concept of regional innovation systems (RIS) has acquired increasing popularity, especially given the socio-economic challenges introduced by the rapidly globalizing economy, such as increasing international competition and hybridization of markets, as well as the apparent shortcomings of

F. Bertolotti (✉) · E. Mattarelli · P. Ungureanu
University of Modena and Reggio Emilia, Reggio Emilia, Italy
e-mail: fabiola.bertolotti@unimore.it

© Springer International Publishing AG, part of Springer Nature 2019
M. Granieri and A. Basso (eds.), *Capacity Building in Technology Transfer*,
SxI – Springer for Innovation / SxI – Springer per l'Innovazione 14,
https://doi.org/10.1007/978-3-319-91461-9_4

traditional regional development models and policies (Asheim and Isaksen 2002; Cooke et al. 1997; Doloreux 2003, Doloreux and Parto 2005; Etzkowitz and Leydesdorff 2000; Leydesdorff and Etzkowitz 1996).

To this regard, technology transfer (TT) is deemed a vehicle for the development of RIS (Lazzeroni and Piccaluga 2003). Therefore, answering questions about why and how certain regional systems are more successful than others in TT becomes paramount. Within this framework, TTOs acquire a prominent position given their role as central hub in the web of relationships between university, industry, government, firms, and other stakeholders (Benassi and Di Minin 2009; Bigliardi et al. 2015; Clarysse et al. 2014; Cooke and Schienstock 2000; Geuna and Muscio 2009; Roxas et al. 2011). Their capacity and performance become instrumental in RIS's success and prosperity.

Two fields of research commonly address these issues. On one hand, we find a growing body of literature on systems of innovations, clusters, and innovative milieu (Asheim et al. 2011). These studies include multiple key topics such as the typologies and varieties of RIS, their impact and efficiency in terms of establishing regional competitive advantages, the boundaries between RIS and other systems such as clusters, networks, regions or institutional fields, or the role of knowledge transfer and learning, and human capital acquisition (Asheim and Isaksen 2002; Amin and Thrift 1995; Cooke and Schienstock 2000; Doloreux and Parto 2005; Edquist 1997; Laranja et al. 2008). On the other hand, studies devoted to understanding the factors able to influence the success and performance of TTOs pay great attention to the pivotal role of the environment in which TTOs operate. For instance, the fourth dimension of the CCODE model (Frederick and Granieri 2015), specifically developed to suit the growth needs of TTOs, refers to the environment (stakeholders) as a critical growth driver. Extant research enriches our comprehension of the impact of the location of TTOs in a specific geographical region (e.g. Cooke 2002), the association with a specific university (e.g. O'Shea et al. 2005), and the maturity and strength of the relationships that single TTOs establish with the broader business community and financial stakeholders to create a supportive ecosystem (e.g. York and Ahn 2011). Some chapters of this book provide excellent reviews of these topics and therefore a deeper treatment is out of our scope.

Despite the constant expansion of research on RIS and on the success factors of TTOs as brokers within RIS, the internal dynamics of RIS, and their impact on how a RIS evolves through time have remained relatively overlooked. For instance, we know little of how partners' interactions shape different governance arrangements or business models inside the RIS, and even less do we know about what pushes partners to change the organization of a RIS through time, and what consequences that can have on how the system evolves and the performance of TTOs as brokers. However, understanding these organizational dynamics is of paramount importance because regional systems of innovation are experiencing increasing environmental turbulence and are often not as successful as desired because of the inability of partners to govern instability and change.

Moreover, studies interested in the interactions of TTOs with their ecosystems often adopted a linear perspective, focusing on the supply side (innovation/IPR generation requires high quality research to motivate industry to be part of the TT process) and the delivery side (a weak ecosystem not able to absorb research has a negative influence on the performance of TTO).

Drawing on the literature on inter-organizational and hybrid partnerships (Austin 2010; Austin and Seitanidi 2012a, b; Bryson et al. 2006; Geddes 2008; Gray 1989; Koopenjan and Klijn 2004; Le Ber and Branzei 2010; Selsky and Parker 2005) we put forth a process-based perspective on the evolution of regional innovation systems. Specifically, we aim at theorizing on how perceptions of environmental turbulence may influence partners' decisions to adopt a given organization model for the RIS (and in particular, for the broker/TTO that manages it) and on the decision biases that perceptions of turbulence may generate at the individual and interpersonal level, with a particular focus on the consequences of decision biases on the evolution of the organizational structure of the TTO through time. We suggest that perceptions of turbulence, and the urgency for action that they trigger inside regional innovation systems, can play an important part in explaining why complex innovation partnerships often deviate from the intended direction, and produce outcomes that are difficult to judge, or come across negatives frames of public opinion. Studying the relation between perceived turbulence, TTO organizational structure/business model, and decision biases at the RIS partnership level, implies adopting a process-based perspective on the evolution of hybrid partnerships in general (Austin 2000; Austin and Seitanidi 2012b; Le Ber and Branzei 2010; Seitanidi and Lindgreen 2010) and on the evolution of regional evolution systems, in particular (Iammarino 2005; Rip 2002). We argue that the VUCA model created by Bennett and Lemoine (2014a) can constitute a useful tool for analyzing turbulent environments through four interrelated elements: volatility, uncertainty, complexity and ambiguity. Importantly, we will use the VUCA model to analyze different configurations of turbulence that hybrid partnerships experience throughout their lifecycles. For this purpose, we will make a distinction between emergent models and structured models of a TTO in RIS partnership.

The rest of the chapter is organized as follow. We first define RIS and role of brokers, such as TTOs. We then present the issue of turbulence and introduce the VUCA model to analyze it. We theorize about how the perceptions of VUCA environments influence the evolution of partnerships through time—in particular its governance models. We also offer insights on the possible decision making-pathologies that might occur among partners in the different evolution stages.

In our conclusions it will be clear how the analysis conducted with the VUCA model can become useful in designing more sound and conscious capacity building actions for technology transfer operations.

2 RIS in a Changing World

We know that the triple helix model (structuration of university–industry–government relations) is actually constituted of plethora of varying institutional arrangements between these actors (as well as other actors that may add up to the model,
depending on case and sector of reference), including different organizational structures, different choices of intermediation, varying degrees of autonomy and decision
centralization, as well as differing patterns of communication inside the partnership
(Asheim and Isaksen 2002; Cooke et al. 1997; Cooke and Schienstock 2000;
Doloreux and Parto 2005; Etzkowitz and Leydesdorff 2000; Laranja et al. 2008).

Two interconnected aspects can be considered to understand how RIS partnerships evolve throughout their lifecycle. The first is their constant need for change,
and the endless transition towards business models that can answer more promptly
the environmental challenges that participants face individually and collectively (as
a partnership). Changes in technological innovation, new conditions of competition
and boundary switching in markets and prevailing sectors, and changes in institutional regulations are just some examples of the environmental challenges that
require RIS to pursue continuous change. Innovation, in particular, requires changes
in both the innovators and the systems they have put in place in order to innovate.
However, when partners must make decisions about the RIS throughout the
innovation processes, many challenges arise. As partners become both observers,
participants and decision makers in their environments, their perceptions of how
affording or constraining the RIS environment is, may significantly impact how
they decide to structure their collaboration. The second related aspect, we argue, is
how participants in a RIS decide to use their broker/TTO to structure their
collaboration.

3 The Role of RIS Brokers for RIS Evolution

The difficulties that can arise before, during and after the formation of cross-sector
partnerships necessitate the involvement of partnership experts who can convene
partners and support them during the partnering process (Seitanidi and Crane 2014;
Kolk et al. 2008; Huxham and Vangen 2013; Selsky and Parker 2005; Stadtler and
Probst 2012; Waddell and Brown 1997). In RIS partnerships (including licensing
arrangements and R&D consortia), these roles are usually occupied by TTOs, which
practitioners and academics alike generally consider of great importance for successful partnerships (Benassi and Di Minin 2009; Bigliardi et al. 2015; Perez and
Sánchez 2003; Roxas et al. 2011; Villani et al. 2017). Studies suggest that such
figures play an important part in accompanying RIS throughout their development
stages. For instance, since cooperation and learning behaviors do not emerge
spontaneously, it is necessary to support interaction around complex issues in
network-settings. Studies have generally referred to these support strategies as network management (Agranoff and McGuire 2003; Clarysse et al. 2014; Geuna and

Muscio 2009; Huggins 2008; Siegel et al. 2007). Accordingly, managing complex cross-sector networks such as the ones that constitute regional innovation systems is a challenging mission, with managers often having difficulties in gaining partners' legitimation and support (see also Koopenjan and Klijn 2004). Furthermore, acting as RIS brokers requires numerous skills such as metaknowledge of the network and conflict negotiation skills, given that TTOs (just as RIS managers and brokers in general) are required to act in situations where well-defined hierarchical relations are lacking. As a matter of fact, the network manager has the role of interaction mediator and stimulator: It can be a person, or an entire organization. In the second case, its structure can be more or less centralized, more or less autonomous, more or less specialized with respect to the regional innovation system. Additionally, the broker role can be even fulfilled by more than one actor at the same time, both public and private, and its agenda can be more or less well-defined. Moreover, brokers present a moving target. They can be engaged in a large number of activities for many different stakeholders under very diverse circumstances (Cooke et al. 2004; Kallio et al. 2010; Kauffeld-Monz and Fritsch 2013; Klerkx and Leeuwis 2009; Geuna and Muscio 2009; Huggins 2008; Siegel et al. 2007; Stadtler and Probst 2012). Complementarily, the specific literature on TTO underlines that a common challenge they often face is related to the mandate they receive from their respective PRO (Bruneel et al. 2010; Muscio 2010; Stadtler and Probst 2012; Vurro et al. 2010). Mandates, at times, are not only badly articulated, but they also lack strategic aims or they focus too much on miscellaneous activities not directly related to technology transfer actions, with the likely consequence of wasting scarce resources and impairing TTO's chance of developing useful technology transfer capacities (this aspect is more specifically dealt with under chapter 4 of this book). However, little is known about how the organization of TTOs reflects the status quo and the evolution of a RIS partnership. Furthermore, there is almost no evidence about the relationship between perceived environmental turbulence, decisions about how to organize the RIS, decisions about the role assigned to the RIS brokers, and the evolution patterns of the RIS itself. As follows, we first turn to a discussion of perceived turbulence in cross-sector partnerships, and then try to connect it to decisions that partners make about the organizational structure/model of the RIS.

4 Environmental Turbulence and RIS Partnerships

Cross-sector partnerships for innovation are on the rise because organizations 'want something from each other' as they can no longer fulfil their task alone, whether they like to or not. Already more than 40 years ago Emery and Trist (1965) argued that increased environmental complexity triggers the perception of organizations from both the private and the public sectors that the "ground is in motion" (Koopenjan and Klijn 2004). According to the seminal study of Bryson et al. (2006), all interorganizational relationships, cross-sector collaborations included, are more likely to form in turbulent environments (see also Austin and Seitanidi

2012a; Geddes 2008; Gray 1989; Koopenjan and Klijn 2004; Selsky and Parker 2005). In particular, the formation and sustainability of cross-sector collaborations are affected by driving and constraining forces in their competitive and institutional environments. The complex, uncertain and ambiguous nature of many societal problems can be connected to a number of characteristics of the network society (Castells 2000; Bryson et al. 2009; Koopenjan and Klijn 2004; Selsky and Parker 2005). From such standpoint, private companies increasingly operate in a world-wide theatre, while economic activities are less bound to geographical locations and territories, and the idea that their main objective is profit maximization is less frequently accepted. As a consequence, private organizations see community needs as opportunities to develop and demonstrate new business technologies or to serve new markets that are out of their usual reach, as well as enter conversations where private initiative is commonly viewed with skepticism (Fontana et al. 2006; Leydesdorff and Meyer 2006; Mohnen and Hoareau 2003). As far as governments are concerned, visions of what the general interest is, are less commonly shared than used to be. Whether they are looking to expand jurisdiction to new economic sectors, retrieve skills and competencies that are not available internally, achieve legitimation in new markets and fields, or simply adjust to the requirements of the New Public Management, public organizations increasingly establish collaborations with other sectors (Osborne 2010; Skelcher 2005).

According to Koopenjan and Klijn, (2004), the problems that cross-sector partnerships come together to deal with, are characterized by a high degree of wickedness. From such standpoint, involved parties disagree not only about the solution, but even about the very nature of the problem. The fact that governments, businesses and civil society are unable to find solutions individually is associated to an increasing degree of perceived turbulence in their environments, such as complexity and uncertainty. It thus appears that environmental turbulence prepares the arenas of cross-sector interactions. However, little is known about how this occurs exactly. We propose that Bennett and Lemoine's (2014) conceptualization of organizational performance in VUCA environments can be expanded and used as a canvass to systematize the challenges and opportunities of collaboration for RIS development and to investigate the characteristics of the ecosystems in which TTOs operate. As follows, we first illustrate the assumptions of the VUCA frame as conceived by the authors Bennett and Lemoine (2014) at the organizational level and expand them through the connection with evidence from studies at the interorganizational level.

5   A VUCA Approach to the Challenges and Opportunities of Partnerships for RIS

Bennett and Lemoine's (2014a) model proposes that four characteristics of the environment pose serious threats to the performance of organizations. These characteristics are volatility, uncertainty, complexity and ambiguity (VUCA).

According to the authors, the analysis of their distinctiveness as well as their interactions is necessary to empower decision makers and place them in the condition to allocate resources and preserve organizational performance.

According to the VUCA frame, *volatility* occurs when environmental changes are frequent and unstable so to make them unpredictable. Future conditions of the organization are thus difficult to ascertain. While organizations may enroll in partnerships with actors of their eco-systems to face environmental volatility, we argue that partnerships themselves can be threatened by volatility (Doz and Hamel 1998; Selsky and Parker 2010; Van Marrewijk et al. 2008). Even if partners are aware of the heterogeneous competencies, interests, and actions of the other partnering organizations, they have difficulties in monitoring them because they often change course (see Austin 2000; Babiak and Thibault 2009). To cope with volatility, agility should be developed through the creation of slack resources that in a partnership could help partners adapt to volatile changes inside or outside the partnership.

Uncertainty describes conditions where the lack of adequate information and knowledge makes hard to understand if certain events could be significant; thus, it becomes difficult for an organization to prepare and respond to them effectively. If we move at the inter-organizational level, partners, tasks, and technological uncertainty strongly influence the governance model that partners choose for a partnership, which in turn impacts its ability to grow (Cooke 2002; Koppenjan and Klijn 2004; Santoro and McGill 2005). The uncertainties that partners experience are a function of their meanings, roles, and relationships with the other partners (see Bennett and Lemoine 2014; Brugnach et al. 2008). Increasing the sources of information is critical to reduce uncertainty (Bennett and Lemoine 2014). Organizations should move beyond existing information sources to gather new data about the other partners and consider it from new perspectives.

Complexity refers to situations where the multiplicity of elements and interconnected parts is so high to make decision-makers overwhelmed by the task of collecting and understanding relevant information. If we move at the level of interorganizational collaborations, complexity can be one of the main causes for their failure because it directly relates to difficulties of coordination of many independent organizations each characterized by different interests, cultures and business models. For instance, partnerships involving numerous and different actors can easily go off track in the attempt to align long-term objectives of the partnership and short terms interests of partners or to align goals and activities of the partnership and goals and activities of each partner (Googins and Rochlin 2000; Le Ber and Branzei 2010). Dealing with complexity in a hybrid partnership usually implies negotiating organizational goals, operations and processes to match those of the other partners, or reducing them to a set of common goals which are easier to implement (Huxham and Vangen 2000a, b).

Last, *ambiguity* characterizes circumstances that because of their newness make it difficult to understand the nature of cause-effect relationships (Bennett and Lemoine 2014). Ambiguity shares with uncertainty the lack of clarity of information but it also includes: (1) lack of clarity about the importance of environmental variables; (2) lack of clarity about cause-effect relationships between variables; and (3) confusion about available courses of action and their potential effects (Carson et al. 2006; Huxham and Vangen 2000a, b). Consequently, because of ambiguity, organizations can hold different perceptions of the same environment and of the decisions and actions that should be taken next (Carson et al. 2006; Huxham and Vangen 2000a, b). Within a partnership, ambiguity can further propagate heterogeneity and prevent partners from developing a common ground for successful interaction. Experimentation is considered by Bennett and Lemoine (2014) a best practice to reduce ambiguity. Organizations can clarify the common goal and determine what strategies are beneficial or detrimental in new situations through intelligent trials and errors.

Recently, it has been also suggested that partnerships, in the attempt to manage VUCA environment, might introduce new sources of VUCA (Sydow et al. 2013). For instance, when a partner decides to lower partnership uncertainty by gathering information about the other partners, the agility of the partnership, and thus its ability to cope with a volatile environment, might also suffer. Therefore, it becomes paramount for each partnership, and for each partnering organization, to identify the best practices for reinforcing virtuous cycles while inhibiting vicious cycles. Keeping track of how VUCA environments influence the evolution of partnerships through time is the first step in this endeavor.

6 The Relationship Between VUCA Perceptions and Governance Models of RIS and TTOs

Studies have shown that cross-sector collaborations are more likely to succeed when they establish—with both internal and external stakeholders—a governance model which results in a (more or less) separate entity (e.g., independent/subordinate TTO), provided the entity becomes a protected and trusted arena for interaction among members.

An elusive question is what constitutes governance for networks or collaborations (Austin and Seitanidi 2011; Bryson et al. 2006). If we assume that networks are horizontal systems, then a hierarchical concept such as governance can be troublesome (Provan and Kenis 2005). However, governance as a set of coordinating and monitoring activities must occur in order for collaborations to survive, regardless of how autonomous partnering organizations are.

The structure of the governance model is influenced by the initial conditions of a partnership and a key process in cross-sector collaboration is negotiating formal and informal agreements about the purpose of the collaboration after some initial agreement on problem definition has been reached. This is usually done by considering structural aspects such as roles, responsibilities, and decision-making authority (Austin and Seitanidi 2012b; Huxham 2000a, b, 2013; Le Ber and Branzei 2010; Sharfman et al. 1991). Moreover, collaboration scholars suggest that structure is often influenced by context. For example, changes in government policy often destabilize RIS systems, leading to a perceived need to rearrange the structure of ties among members (Sharfman et al. 1991; Stone 2004). The same TTOs are often destabilized when they face changes in the leadership (and the related TT policies) of the universities to which they are attached. This is something to be kept in mind when considering the timing of a capacity building action and its intended consequences as it may suffer from organizational changes of the TTO due to management turnover.

System stability and the degree of resource munificence (Austin and Seitanidi 2011; Bryson et al. 2006; Human and Provan 1997; Provan and Milward 1995; Sharfman et al. 1991) have also been called in question as initial conditions that explain how a partnership evolves. The strategic purpose of a partnership also appears to affect partners' decisions about the most appropriate structure (see also Agranoff and McGuire 1998; Austin 2000; Klijn 2008; Koopenjan and Klijn 2014; Westley and Vredenburg 1991).

Many studies have referred to partners' choices between emergent and highly structured models. The first approach argues that a clear understanding of mission, goals, roles, and action steps is more likely to emerge over time as conversations between individuals, groups, and organizations grow larger and deeper. This approach, which Mintzberg et al. (1998) have famously called 'emergent' seems most likely when collaboration is not mandated (for discussions see Austin and Seitanidi 2011; Bryson et al. 2006; Huxham and Vangen 2005, Winer and Ray 1994). The second model, by contrast, requires a careful ex-ante articulation of mission, goals, objectives, roles and responsibilities, and phases or steps (Mattessich et al. 2001). This approach—which Mintzberg et al. (1998) have called 'deliberate'—is often cited as an important key to success in complex partnerships and appears to have a higher likelihood of being implemented when collaboration is mandated (for a discussion see Bryson et al. 2006). Contingencies such as network size and the degree of trust among members influence which form is appropriate.

However, more recent studies have suggested that partners rarely choose one form or the other (Seitanidi et al. 2010; Gray 1996). For instance, there is evidence that cross-sector collaborations are more likely to succeed when they combine deliberate and emergent planning. Additionally, it has been argued that models change as a partnership evolves throughout its lifecycles and is confronted with different internal and external (i.e., environmental) challenges. For instance, Bryson et al. (2006) have suggested that organizations choose the most appropriate governance model among a set of alternative choices, such as (1) self-governing structures in which decision making occurs through regular meetings of members or through informal, frequent interactions; (2) decisions to invest in a lead organization that provides major decision-making and coordinating activities; and (3) decisions to create a network administrative organization, which is a separate organization that has the mandate to oversee the partnership's affairs.

We suggest that a greater attention should be paid to VUCA configurations that characterize each model, in the attempt to understand how perceptions of VUCA in one model may push the partnership to adopt the other model. Also, it is important to overcome linear models in which partnerships are described as having progressive trajectories (from emergent to structured models). For instance, there may be cases in which partnerships move back and forth between more emergent and more structured models, according to perceptions of turbulence in their environments. Understanding those aspects is also instrumental for designing appropriate capacity building actions for TT partnerships, particularly as far as TTOs activities and functions evolve towards the supply of brand new services, such as proof-of-concept funding in connection with local investors (Munari and Toschi 2018).

Table 1 summarizes the analysis that we present in the next paragraphs.

Table 1 Governance models of RIO, the influence of VUCA, the role of TTOs, and pathologies

Definition	Governance models of the RIO		
	Emergent	Structured	Administered platform
How VUCA influences the governance model of the RIS	The volatility of the objectives in a hybrid partnership, the uncertainty of the long-term outcomes, and partners' difficulty in disambiguating fluid and divergent interests, might induce partners to opt for an unstructured partnership model	Partnerships become more structured to lower the perceived uncertainty and ambiguity, but experience higher complexity, necessary to deal with elaborate rules, standards and procedures, and to achieve coordination and communication at the partnership level	By acting on their own account, partners may reduce the perceived level of volatility and uncertainty in the partnership
The structure and role of TTO	Partners may look for TTOs that have a flexible, ill-defined organization, at least with respect to the goals and activities of the partnership	Partnering organizations remain focused on their core missions while they delegate TTOs to act as formal brokers and to operate in new fields of interests.	TTOs have a mere support role, identifying synergies between projects and suggesting more efficient recombination across innovation projects
Decision Pathologies of the RIS	Perpetual reconsideration of decisions that creates a state of collective ambivalence preventing partnerships from progressing forward	Delay or postponement of a course of action one intends to pursue because of unwillingness or unreadiness to deal with emergent constraints (at the personal, interpersonal or organizational level)	Circumstances in which commitment to a course of action is reinforced in the face of adverse information about its results, usually by generating sufficient resources of capital and legitimacy to ignore evident proofs of negative performance

6.1   VUCA Perceptions and Emergent Governance Models

Studies suggest that partners adopt emergent partnership models because they seek flexible tools that would allow them to define their involvement in the partnership gradually, as they make sense of the other partners and of the relationship between each other's goals. Specifically, in their initial stages, cross-sector partnerships are filled with volatility, uncertainty and complexity. Uncertainty arises when parties are confronted with societal problems and do not know what sort of effort will allow them to reach the common goal, and even ignore whether the common goals of the partnership are feasible. Koopenjan and Klijn (2004) have proposed that in the emergent stages of a partnership that aims at dealing with complex issues, uncertainty has three manifestations: substantive, strategic, and institutional. Substantive uncertainty concerns the availability of information of how a complex issue can be addressed in an innovative way. Especially in RIS projects, necessary information is often lacking or may not be available in a timely manner. In addition to substantive uncertainty, dealing with ill-defined problems in an innovative way brings along strategic uncertainty at the partnership level. Although actors come together in the name of a shared goal, their strategic agendas often remain separate because their organizational frames are diverse, and their main stakeholders have different concerns and require different forms of accountability, which they carry with them inside the partnership (Le Ber and Branzei 2010; Selsky and Parker 2005; Ungureanu et al. 2018; Vurro et al. 2010). In addition, institutional uncertainty is often an initial condition of hybrid partnership, given that participants not only have different perceptions, objectives, and interests, but they also come from different organizations, administrative levels and networks. In addition to different types of uncertainty, what makes matters volatile in the beginning of a partnership is the fact that actors are rarely aware of each other's unique perceptions and organizational frames. The consequence is that a large variety of auto-referential strategies may develop around a complex issue, causing an overflow of uncoordinated or misaligned courses of action. In turn, partners may have difficulties keeping track of everything that happens inside the partnership. The consequence is the perception of dispersive, insufficient or unreliable information about the other participants in the network. Furthermore, ambiguity is often another condition that encourages partners to adopt an emergent model. Since actors know each other only to a limited extent, they may fear each other's hidden interests and feel threatened by potentially divergent perspectives. Since the communication structures are usually also underdeveloped in the initial stages, partners may not have the appropriate tools to disambiguate the contradictory signals coming from others. Because of these mechanisms, it is difficult to predict what strategies actors will choose and how the 'interaction' of these strategies will influence the problem situation and problem-solving process (Koopenjan and Klijn 2004). As such, choosing a strategy that allows partners to remain loosely-coupled can be reassuring, as it gives partners the necessary time to understand how to make sense of misalignments, contradictions and uncertainties in the network.

6.2 The Role and Structure of TTOs in Emergent Governance Models

The above analysis shows that the volatility of the objectives in a hybrid partnership, the uncertainty of the long-term outcomes, and partners' difficulty in disambiguating fluid and divergent interests, might induce partners to opt for an unstructured partnership model. This condition is also likely to reflect in the organization of the broker/TTO, as partners may look for brokers that also have a flexible, ill-defined organization, at least with respect to the goals and activities of the partnership.[1] In a complex network where hierarchical structures are missing, and unexpected strategic turns are an intrinsic characteristic of interaction processes, partners may feel motivated to avoid making strong and explicit commitments for as longer as possible. The emergent model's main advantage is keeping boundaries between single partners', brokers' and partnership's interests highly mobile and permeable. An open and flexible collaboration model allows partners to gradually acclimatize to working together, without worrying too much about the uncertainties of partnership evolution, about the ambiguous intentions of the other partners, or about the volatility of partnership events. Freedom to make experiments and the possibility to choose when and how much to invest in the partnership, also encourage partners to search for a flexible, unstructured type of brokerage that will not push towards anticipated commitments. A broker that does not require partners to expose themselves right away on pressuring matters of which only limited information subsists, can provide partners with a sense of reassurance and protection against uncontrollable, rapidly changing and highly undetermined types of interactions. Partners are shown to seek such relations actively in the first stages of their collaboration (Austin and Seitanidi 2012; Vurro et al. 2010)

However, structuring the partnership according to an emergent model, identifying/creating brokers that have a similar orientation, can result in a high degree of uncertainty about how the process will be handled and how the interaction with other actors will develop (Koopenjan and Klijn 2004). Thus, by adopting an emergent model, the initial perceptions of substantive, strategic and institutional uncertainty may have an opposite effect—i.e., increase perceptions of environmental turbulence, instead of decreasing them.

As follows, we provide an explanation of the individual and inter-personal biases that push partners to adopt partnership solutions that may end up having the opposite effect than the one initially envisioned.

---

[1]It is interesting to notice that this search for flexibility could clash with the limited discretionality afforded to TTOs and this could limit their capacity to develop certain functions toward the environment.

6.3 Possible Decision-Making Pathologies in Emergent Models: Indecision

When confronted with high levels of perceived VUCA, partners may lean toward indecision: a pathological condition of perpetual reconsideration of decisions that creates a state of collective ambivalence preventing projects to progress forward (Denis et al. 2011).

In the literature on interorganizational collaboration, Huxham (1996) has referred to this phenomenon as collaboration inertia: the tendency to slow down and oscillate in accomplishing the courses of action for which the partnership was established. Indecision in partnerships derives from the inherently collective, heterogeneous and politically vested interests inside partnerships which keep the decision-making process continuously suspended in a state of uncertainty. Situations of reactive leadership, uncertainty about available resources, and long-term horizons have been associated with indecision pathologies in interorganizational collaboration (see also Huxham 1996). In other words, indecisions seem to stem from the conditions of an emergent model of organization. For example, psychology studies have connected indecision to perceived uncertainty, arguing that the more actors feel unable to say which action will produce which outcome, the more they remain stuck in alternative scenarios and feel unable to show clear preference for either of them. Lack of information, valuation difficulty, and outcome uncertainty, are said to be the pillars of indecision biases (see Anderson 2003; Rassin 2007).

For instance, as soon as partners adopt an emergent model, ambiguity arises from many features of a free, fluid and flexible membership, including perceptions of who belongs to the partnership, what the other members actually represent (e.g., themselves, their organizations, or a particular identity group), what their next moves will be, and how to understand and deal with turnover among members (e.g. top elected officials leave or alter their level of involvement in the collaboration, Crosby and Bryson 2005; Kastan 2000).

As Denis et al. (2011) and Gray (1996) suggest, indecision may lead to conflict, as each partner puts the blame on the other partners for the lack of progress in the partnership or tries to exercise influence and resource control over them (see also Babiak and Thibault 2009). Conflict may be further exacerbated when the collaborating organizations differ in status (either because of size, funding, or reputation) (Merrill-Sands and Sheridan 1996).

To deal with threats such as partnership paralysis and conflict, partners may be tempted to propose more structured and centralized models, which swing the pendulum to the side of the "deliberate model".

6.4 VUCA Perceptions and Centralized Governance Models

Partners commonly transition from an emergent model to a deliberate, more structured model, to deal with the additional VUCA threats that an emergent model may generate. To unblock a partnership that is pervaded by indecision, partners may invest in a broker's ability to collect information. This response is based on the idea that volatility and substantive uncertainty are a result of lack of information or knowledge. Centralizing information collection in the hand of a broker can be interpreted as a way to increase the quality and objectivity of research and to reduce differences in interpretation (Koopenjan and Klijn 2004; Vurro et al. 2010). The more structured a partnership becomes, the lower the uncertainty and ambiguity, and the higher the complexity necessary to deal with elaborate rules, standards and procedures, and to achieve coordination and a functional communication flow at the partnership level (Austin 2000; Austin and Seitanidi 2012b; Sharfman et al. 1991; Selsky and Parker 2005; Vurro et al. 2010).

It is important to highlight that structuration may occur differently inside partnerships. For instance, Bryson et al. (2006) distinguish between decisions to invest in a broker (i.e., lead organization) that allows for centralization in decision-making and coordinating activities, and decisions to create an administrative organization to run a partnership platform or network, which implies higher integration but also less delegation and centralization.

6.5 The Role and Structure of TTOs in Centralized Governance Models

Centralization may prove a crucial indication for the role of a TTO in structuring a partnership at RIS level, since TTOs within regional universities and research centers often times are perceived as pivotal.

The adoption of a centralized brokering model can generate a sort of ambidexterity in the partnership such that partnering organizations remain focused on their core missions while they delegate boundary organizations to operate in new fields of interests. The centralized brokering model is intended as a way to face volatility, by making sure that specific roles and responsibilities are created inside the partnership in order to keep track, collect, and diffuse information regarding the single partners in a timely and efficient manner.

However, the same studies that show why centralized brokering models are adopted, also explain why most of the times they fail to produce the desired outcomes at the partnership level. Specifically, studies show that in practice, setting up a more structured information gathering system rarely lead to a decrease in volatility and uncertainty. The broker/TTO's search for information rarely solves the issues related to the existence of different opinions about a problem. As such, information available may further increase conflict inside a partnership, as partners try to use it to support their own interests. Quite frequently, vehement discussions emerge about the validity of information gathered by the broker, or about the validity of the broker itself (Stadtler and Probst 2012). In turn, this undermines the clarity of roles and responsibilities that the higher structuration was expected to bring along, causing additional volatility, uncertainty, ambiguity, and at times also complexity inside a partnership. Additionally, by centralizing decision making in the hands of a broker, partners underestimate their external dependencies and adopt a limited, self-reassuring approach to the nature of the situation and the interests involved. Most importantly, it must be noted that the transition from an emergent to a highly centralized type of partnership can create a paradoxical situation by which partners formally delegate decision making and information gathering to a brokering organization, but informally undermine the brokers' authority and try to regain centrality in the partnership, hoping to succeed in pushing through their favorite solution.

We suggest that the main risk of a centralized brokering model has to do with partners' unrealistic attempts to switch from a model with high perceived VUCA threats (specifically, VUA threats) to a model that is theoretically low on VUCA (it delegates the sources of environment turbulence to a third party) but in practice can prove to be even more VUCA-pervasive than the former. The will to delegate responsibility to a broker, on the one hand, and the fear to let go off control, on the other, may generate paradoxical situations inside the partnership and lead to opposite effects from the ones envisioned (e.g., split hierarchy, double unaccountability, overflowing goals). While we do not imply that there are no examples of successful brokering model dealing with complex multi-party innovation projects, we suggest that the likelihood that brokering organizations reach the objectives for which they were created decreases as the number and heterogeneity of goals assigned to these brokers increase, especially when partners are reluctant to hand over control over partnership projects. Given the limited evidence on this topic, future research is needed to investigate the conditions in which the transition from an open model to a delegation model generates negative, rather than positive consequences for the evolution of the partnership.

Summarizing, we have suggested that when coming from an unstructured partnership model, the adoption of a centralized brokering model can have the opposite consequence than the one for which it is designed, and thus increase perceptions of partnership turbulence, instead of decreasing them. Also in this case the mechanism can be explained through a decision pathology that we refer to as procrastination.

6.6 Possible Decision-Making Pathologies in Centralized Models: Procrastination

Procrastination refers to the delay or postponement of a course of action one intends to pursue (Akerlof 1991; Harris and Sutton 1983). Studies suggest that organizations engaged in complex projects might push in the future those courses of action that are less straightforward or that they feel less prepared to deal with, such as clarifying broad or ambiguous project goals, generating agreement, solving conflicts or negotiating the interests of different parties (Van Marrewijk et al. 2008; Flyvbjerg et al. 2005). Procrastination is more likely when time horizons are long (Schouwenburg and Groenewoud 2001; Strongman and Burt 2000) and when actors confront with a strong and stable task averseness—i.e., they must delegate decision making to a broker, but they do not wish to do so because they believe that delegation might bring about negative consequences for their own interests (see Van Marrewijk et al. 2008).

What is interesting to notice is that procrastination can lead to higher levels of VUCA. When partners promise to delegate and legitimate a broker but fail to do so because they fear losing control, they may end up further losing their trust in the other and in the mission of the partnership overall (Sharfman et al. 1991). Also, the designated broker's authority risks to be seriously undermined, as partners cast additional shadows on the utility of the information that the broker gathers and releases within the partnership, or on their ability to act in the best interests of each partner. As roles and responsibilities are questioned, ambiguity is generated. As informational validity is questioned, uncertainty may rise. In addition, as formally centralized decision-making is undermined by an informal system based on emergence and decentralization, volatility may rise even to higher levels than in the emergent stage.

6.7   VUCA Perceptions and Administered Platform/Network Models

Consistent with Bryson et al. (2006), in more mature stages, partners may either adopt a 'platform' model in response to the high VUCA of the emergent model or to deal with the VUCA triggered by the centralized brokering model. This model commonly tries to restore a fluid organization model similar to the one in the first phase; It entails more systematic interactions that aim at providing high integration with low centralization. From the perspective of capacity building, the platform model is likely to shed more light on the process by which TTOs are created at local levels or shared facilities developed in order to increase the intensity of technology transfer to market.

An administered platform model overcomes the problems of the brokering model because it gives space to public institutions, businesses and other social institutions to interact directly, without intermediation. While in the brokering model partners assigned to the broker a general mission of innovation and technology transfer which the broker generally articulates and implements according to its own means, in the administered platform model a higher number of actors plays a more active role. In an open platform, the boundaries between shareholders and stakeholders (i.e., interested individuals, organizations or communities that are situated outside the partnership) become blurred. The partnership often becomes a constellation of small autonomous projects that plan, finance, implement, promote and monitor interests of direct concern such that partnering organizations must act as entrepreneurs in their own projects (Austin 2000; Bryson et al. 2006; Koch and Buser 2006; Selsky and Parker 2005, 2010). By acting on their own account, partners may reduce the perceived level of volatility and uncertainty in the partnership. The high degree of ownership that the partners maintain in this model may encourage them to invest in the partnership more, because the connection between the partnership goals and their own goals becomes more straightforward.

6.8 The Role and Structure of TTOs in Administered Platform/Network Models

While in the centralized brokering model the brokering organization is responsible for finding resources and assembling them around specific shareholder interests, in the platform model the boundary organization has a mere support role, identifying synergies between projects and suggesting more efficient recombination across innovation projects (Rein and Stott 2009; Klerkx and Leeuwis 2009). To give some examples, in a platform model, the broker organization can guarantee the rationalization of the activities hosted on the platform, creating communication and reporting infrastructures inside and across units of innovation. While this may seem a downgrade of the broker's role, it actually acknowledges more realistically what a designated organization can do for a high number of partners which not only continue to mobilize divergent goals but also are often unwilling to hand over the control. When considering clusterization of TTOs as a way to improve capacity building and create critical masses, the dynamics of the platform model must be carefully considered as they can provide a powerful explanation of internal processes and warn about possible hurdles and sources of failure.

Notably, also the platform model is not without flaws and risks. The main risk is that it increases the complexity and the volatility of the partnership. When partners are not ready to deal with the increasing complexity generated by a platform functioning, the same problems from the emergent model may occur. Also, the effective functioning of a platform model entails an elevated level of volatility because it depends heavily on how much partnering organizations actually behave like entrepreneurs, i.e., the extent to which they take initiative and draw on highly specific and fast-moving information that usually other partners do not have (Cooke et al. 1997; Etkowitz 2008; Fontana et al. 2006). In the absence of these conditions, the RIS platform becomes less dense, and the ability to mobilize network externalities also becomes more modest. While these two aspects have been discussed in the regional innovation literature (see Cooke et al. 2004; Etkowitz 2008), we suggest that future research also on TTOs' clusters (a trend that is somehow taking place in Europe right now, at least in some areas) needs to focus on the interplay between what partners intend by adopting a given model for the partnership, and the unintended consequences that derive from within.

Specifically, platforms are highly complex to manage because they entail actors with diverse backgrounds and partially divergent objectives. The management of such units can provide significant challenges for those who are expected to act as facilitators. Although this last aspect has received less attention both in research and in practice, it is particularly important to set up appropriate infrastructures such as training offers, policy guidelines or access to existing best practices to support the professionalization of platform managers (see also Stadtler and Probst 2012). Significantly, we here suggest that the platform model can also make things worse by trying to make them better. In particular, since platform models require high levels of entrepreneurial initiative and high investments to get single projects going, when outcomes do not keep up with expectations, partners may feel too attached to their initiatives to admit failure and try to actively correct them through joint strategies. We discuss escalation of commitment as a possible bias of partnerships that adopt a platform model.

6.9 Possible Decision-Making Pathologies in Platform/Network Models: *Escalation of Commitment*

Escalation of commitment reflects circumstances in which individuals tend to reinforce commitment to a course of action in the face of adverse information about its results, usually by generating sufficient resources of capital and legitimacy to ignore evident proofs of negative performance. Numerous explanations for why people engage in escalation behavior have been offered, such as refusal to admit having engaged in erroneous courses of action (Brockner and Rubin 1985), a sense of personal responsibility for the initial decision that led to the failing course of action (Staw 1976), the perception of project sunk costs (Arkes and Blumer 1985), or the compelling need to bring a project to completion (Conlon and Garland 1993).

A limited number of contributions have brought evidence that organizations, like individuals, also have problems in terminating commitments, although somehow differently (see Brockner 1992 and Sleesman et al. 2012, for reviews). The typical assumption in the organizational escalation literature is that, when a project does not take off as expected, organizations must choose between withdrawing from a losing project and staying on board to achieve, at least in part, the goals for which it was set up (Brockner 1992; Ross and Staw 1993). It is interesting to notice that when projects are multi-party, have broad goals and a long temporal horizon, such as in hybrid partnerships for regional innovation, escalation seems more likely to occur (see Flyvbjerg et al. 2003; Winch 2013). Specifically, as knowledge grows and complexifies within a platform network, and as initiatives become more concrete and require higher integration, partners become aware of the complexities that surround the platform and new knowledge questions emerge. Specialization according to projects also results in a high degree of fragmentation of knowledge and methods (Koopenjan and Klijn 2004; Bryson et al. 2006, 2009; Nooteboom 2000). As a consequence, partners may feel overwhelmed by the increasing complexity in the partnership, and the volatility of a networked organization, and may either step out of the partnership commitments or, oppositely, engage in escalation of commitment. Future research is necessary to understand when escalation or giving up are more likely to occur. Psychology studies on escalation of commitment suggest that escalation is more likely to occur in contexts in which there is no clear information about opportunities and costs of staying aboard versus giving up, when partners manifest high initiative and high propensity to risk, when there is some available information that partners may benefit from bringing a project to completion. Sunk costs in terms of invested time and resources also seem to play an important role, just as social process, such as public evaluation of the project, strong group identity and conformism to the mission and values of the partnership (for a review see Sleesman et al. 2012).

7   Lessons from the VUCA Model for Capacity Building in Technology Transfer

In this chapter, we proposed that the relationship between perceptions of environmental turbulence and evolution of RIS can be better understood by drawing on Bennett and Lemoine's (2014) analysis of environmental turbulence as VUCA, and on the stream of studies concerned with how decision pathologies emerge across the lifecycle stages of cross-sector partnerships in general (Austin and Seitanidi 2012b; Le Ber and Branzei 2010; Sharfman et al. 1991). We have highlighted that both a partnership's conditions (i.e., goals, interests, commitments) and its decision processes will need to change as the partnership evolves (Huxham and Vangen 2000a, b; Waddock 1989) and, as a consequence, so will the coping strategies enacted by partners to deal with perceived VUCA inside and outside the partnership (Bryson et al. 2006; Eden and Huxham 1996; Sharfman et al.1991).

The VUCA approach offers a sophisticated framework to provide insights on the environment surrounding the TTO, within given RIS, particularly when TTOs are involved in process of clusterization or in the creation of other bridging organizations that are supposed to enhance the transfer of technology at local level. Eventually, also in this respect improving the performance of TTOs requires a holistic approach which cannot overlook the context.

Understanding the characteristics of the environment and how they become influential on the TTOs performance is a first and crucial step towards more effective capacity building actions. The VUCA approach applied to the TTO may contribute to a higher level of awareness in policy makers about the determinants of TTOs (un-) success but also about the set of leverages that can be acted upon to improve the framework conditions which are responsible for TTOs performance.

Pointing out possible directions on how to turn the VUCA model into a normative tool for designing capacity building actions is beyond the scope of this chapter. But blending contributions from the organizational literature and inter-organizational partnerships adds an unexpected ingredient to policy recipes for technology transfer as it reinforces the belief that capacity building is a process affected also by external conditions and that TTOs do not live and operate within an empty cabinet for policy makers experiments, but in a bundle of complex relationships that actively concur in TTOs performance.

References

Agranoff R, McGuire M (1998) Multinetwork management: collaboration and the hollow state in local economic policy. J Public Adm Res Theor 8:67–91

Agranoff R, McGuire M (2003) Inside the matrix: integrating the paradigms of intergovernmental and network management. Int J Public Adm 26:1401–1422

Akerlof GA (1991) Procrastination and obedience. Am Econ Rev 81:1–19

Amin A, Thrift N (1995) Globalization, institutions, and regional development in Europe. Oxford University Press, Oxford

Anderson CJ (2003) The psychology of doing nothing: forms of decision avoidance result from reason and emotion. Psychol Bull 129:139

Arkes HR, Blumer C (1985) The psychology of sunk cost. Organ Behav Hum Decis Process 35:124–140

Asheim BT, Isaksen A (2002) Regional innovation systems: the integration of local 'sticky' and global 'ubiquitous' knowledge. J Technol Transf 27:77–86

Asheim BT, Smith HL, Oughton C (2011) Regional innovation systems: theory, empirics and policy. Reg Stud 45:875–891

Austin JE (2000) Strategic collaboration between nonprofits and businesses. Nonprofit Voluntary Sect Q 29(1_suppl):69–97

Austin JE (2010) The collaboration challenge: how nonprofits and businesses succeed through strategic alliances, vol 109. Wiley, Hoboken

Austin JE, Seitanidi M (2011) Value creation in business—nonprofit collaborations

Austin JE, Seitanidi MM (2012a) Collaborative value creation: A review of partnering between nonprofits and businesses: Part I. Value creation spectrum and collaboration stages. Nonprofit Voluntary Sect Q 41:726–758

Austin JE, Seitanidi MM (2012b) Collaborative value creation: A review of partnering between nonprofits and businesses. Part 2: Partnership processes and outcomes. Nonprofit Voluntary Sect Q 41:929–968

Babiak K, Thibault L (2009) Challenges in multiple cross-sector partnerships. Nonprofit Voluntary Sect Q 38:117–143

Benassi M, Di Minin A (2009) Playing in between: patent brokers in markets for technology. R&d Manag 39:68–86

Bennett N, Lemoine GJ (2014) What a difference a word makes: understanding threats to performance in a VUCA world. Bus Horiz 57:311–317

Bigliardi B, Galati F, Marolla G, Verbano C (2015) Factors affecting technology transfer offices' performance in the Italian food context. Technol Anal Strateg Manag 27:361–384

Brockner J (1992) The escalation of commitment to a failing course of action: toward theoretical progress. Acad Manag Rev 17:39–61

Brockner J, Rubin JZ (1985) Entrapment in escalating conflicts. A social psychological analysis. Springer, New York

Brugnach M, Dewulf A, Pahl-Wostl C, Taillieu T (2008) Toward a relational concept of uncertainty: about knowing too little, knowing too differently, and accepting not to know. Ecol Soc 13

Bruneel J, d'Este P, Salter A (2010) Investigating the factors that diminish the barriers to university–industry collaboration. Res Policy 39:858–868

Bryson JM, Crosby BC, Stone MM (2006) The design and implementation of Cross-Sector collaborations: propositions from the literature. Public Adm Rev 66:44–55

Bryson JM, Crosby BC, Bryson JK (2009) Understanding strategic planning and the formulation and implementation of strategic plans as a way of knowing: the contributions of actor-network theory. Int Public Manag J 12:172–207

Carson SJ, Madhok A, Wu T (2006) Uncertainty, opportunism, and governance: the effects of volatility and ambiguity on formal and relational contracting. Acad Manag J 49:1058–1077

Castells M (2000) Toward a sociology of the network society. Contemp Sociol 29:693–699

Clarysse B, Wright M, Bruneel J, Mahajan A (2014) Creating value in ecosystems: crossing the chasm between knowledge and business ecosystems. Res Policy 43:1164–1176

Conlon DE, Garland H (1993) The role of project completion information in resource allocation decisions. Acad Manag J 36:402–413

Cooke P (2002) Knowledge economies: clusters, learning and cooperative advantage. Routledge

Cooke P, Schienstock G (2000) Structural competitiveness and learning regions. Enterp Innov Manag Stud 1:265–280

Cooke P, Uranga MG, Etxebarria G (1997) Regional innovation systems: institutional and organisational dimensions. Res Policy 26:475–491

Cooke P, Heidenreich M, Braczyk HJ (2004) Regional systems of innovation: the role of governance in a globalized world

Crosby BC, Bryson JM (2005) A leadership framework for cross-sector collaboration. Public Manag Rev 7:177–201

Denis JL, Dompierre G, Langley A, Rouleau L (2011) Escalating indecision: between reification and strategic ambiguity. Organ Sci 22:225–244

Doloreux D, Parto S (2005) Regional innovation systems: current discourse and unresolved issues. Technol Soc 27:133–153

Doloureux D (2003) Regional innovation systems in the periphery. The case of Beauce in Quebec. Int J Innov Manag 7:67–94

Doz YL, Hamel G (1998) Alliance advantage: the art of creating value through partnering. Harvard Business Press

Eden C, Huxham C (1996) Action research for management research. Br J Manag 7:75–86

Edquist C (1997) Systems of innovation: technologies, institutions, and organizations. Psychology Press

Emery FE, Trist EL (1965) The causal texture of organizational environments. Hum Rel 18: 21–32

Etzkowitz H (2008) The triple helix: university–industry–government innovation in action. Routledge, New York

Etzkowitz H, Leydesdorff L (2000) The dynamics of innovation: from National Systems and "Mode 2" to a Triple Helix of university–industry–government relations. Res Policy 29:109–123

Flyvbjerg B, Bruzelius N, Rothengatter W (2003) Megaprojects and risk: an anatomy of ambition. Cambridge University Press, Cambridge

Flyvbjerg B, Skamris Holm MK, Buhl SL (2005) How (in) accurate are demand forecasts in public works projects? The case of transportation. J Am Plan Assoc 71:131–146

Fontana R, Geuna A, Matt M (2006) Factors affecting university–industry R&D projects: the importance of searching, screening and signalling. Res Policy 35:309–323

Frederick P, Granieri M (2015) Development of a holistic tool to identify barriers to success for technology transfer offices (April 20, 2015). Available at SSRN: https://ssrn.com/abstract=2596508

Geddes M (2008) Inter-organizational relationships in local and regional development partnerships. The Oxford handbook of inter-organizational relations

Geuna A, Muscio A (2009) The governance of university knowledge transfer: a critical review of the literature. Minerva 47:93–114

Googins BK, Rochlin SA (2000) Creating the partnership society: understanding the rhetoric and reality of cross-sectoral partnerships. Bus Soc Rev 105:127–144

Gray B (1989) Collaborating: finding common ground for multiparty problems

Gray SF (1996) Modeling the conditional distribution of interest rates as a regime-switching process. J Financ Econ 42:27–62

Harris NN, Sutton RI (1983) Task procrastination in organizations: a framework for research. Hum Relat 36:987–995

Huggins R (2008) The evolution of knowledge clusters: progress and policy. Econ Dev Q 22:277–289

Human SE, Provan KG (1997) An emergent theory of structure and outcomes in small-firm strategic manufacturing networks. Acad Manag J 40:368–403

Huxham C (1996) Advantage or inertia? Making collaboration work. In: The new management reader, pp 238–254

Huxham C, Vangen S (2000a) Leadership in the shaping and implementation of collaboration agendas: how things happen in a (not quite) joined-up world. Acad Manag J 43:1159–1175

Huxham C, Vangen S (2000b) Ambiguity, complexity and dynamics in the membership of collaboration. Hum Relat 53:771–806

Huxham C, Vangen S (2005) Managing to collaborate. Taylor & Francis, Abingdon

Huxham C, Vangen S (2013) Managing to collaborate: the theory and practice of collaborative advantage. Routledge

Iammarino S (2005) An evolutionary integrated view of regional systems of innovation: concepts, measures and historical perspectives. Eur Plan Stud 13:497–519

Kallio A, Harmaakorpi V, Pihkala T (2010) Absorptive capacity and social capital in regional innovation systems: the case of the Lahti region in Finland. Urban Studies 47:303–319

Kastan J (2000) School-based mental health program development: a case study of interorganizational collaboration. J Health Polit Policy Law 25:845–862

Kauffeld-Monz M, Fritsch M (2013) Who are the knowledge brokers in regional systems of innovation? A multi-actor network analysis. Reg Stud 47:669–685

Klerkx L, Leeuwis C (2009) Establishment and embedding of innovation brokers at different innovation system levels: insights from the Dutch agricultural sector. Technol Forecast Soc Chang 76:849–860

Klijn EH (2008) Policy and implementation networks: managing complex interactions. The Oxford handbook of inter-organizational relations, pp 118–146

Koch C, Buser M (2006) Emerging metagovernance as an institutional framework for public private partnership networks in Denmark. Int J Project Manage 24:548–556

Kolk A, Van Tulder R, Kostwinder E (2008) Business and partnerships for development. Eur Manag J 26:262–273

Koopenjan J, Klijn EH (2004) Managing uncertainties in networks. A network approach to problem solving and decision making. Routlegde, London

Laranja M, Uyarra E, Flanagan K (2008) Policies for science, technology and innovation: translating rationales into regional policies in a multi-level setting. Res Policy 37:823–835

Lazzeroni M, Piccaluga A (2003) Towards the entrepreneurial university. Local Economy 18:38–48

Le Ber MJ, Branzei O (2010) (Re) forming strategic cross-sector partnerships relational processes of social innovation. Bus Soc 49:140–172

Leydesdorff L, Etzkowitz H (1996) Emergence of a Triple Helix of university–industry–government relations. Sci Public Policy 23:279–286

Leydesdorff L, Meyer M (2006) Triple Helix indicators of knowledge-based innovation systems: introduction to the special issue. Res Policy 35:1441–1449

Mattessich PW, Murray-Close M, Monsey BR (2001) The wilder collaboration factors inventory: assessing your collaboration's strengths and weaknesses. Wilder Pub. Center

Merrill-Sands D, Sheridan B (1996) Developing and managing collaborative alliances: lessons from a review of the literature. Organ Change Briefing Note 3

Mintzberg H, Ahlstrand B, Lampel J (1998) Strategy Safari: A Guided Tour through the Wilds of Strategic Management, Free Press, New York

Mohnen P, Hoareau C (2003) What type of enterprise forges close links with universities and government labs? Evidence from CIS 2. Manag Decis Econ 24:133–145

Munari F, Toschi L (2018) The intersection between capacity building and finance. In: Granieri M, Basso A (eds) Capacity building in technology transfer. The European experience. Springer, Berlin

Muscio A (2010) What drives the university use of technology transfer offices? Evidence from Italy. J Technol Transf 35:181–202

Nooteboom B (2000) Learning and innovation in organizations and economies. Oxford University Press, Oxford

O'Shea RP, Allen TJ, Chevalier A, Roche F (2005) Entrepreneurial orientation, technology transfer and spinoff performance of US universities. Res Policy 34(7):994–1009

Osborne SP (2010) The new public governance: emerging perspectives on the theory and practice of public governance. Routledge

Perez MP, Sánchez AM (2003) The development of university spin-offs: early dynamics of technology transfer and networking. Technovation 23:823–831

Provan KG, Kenis P (2005) Modes of network governance and implications for public network management, vol 29. In: Eighth National Public Management Research conference, Los Angeles, CA, September

Provan KG, Milward HB (1995) A preliminary theory of interorganizational network effectiveness: a comparative study of four community mental health systems. Adm Sci Q, 1–33

Rassin E (2007) A psychological theory of indecisiveness. Neth J Psychol 63:1–11

Rein M, Stott L (2009) Working together: critical perspectives on six cross-sector partnerships in Southern Africa. J Bus Ethics 90:79–89

Rip A (2002) Regional innovation systems and the advent of strategic science. J Technol Transf 27(1):123–131

Ross J, Staw BM (1993) Organizational escalation and exit: lessons from the Shoreham nuclear power plant. Acad Manag J 36:701–732

Roxas SA, Piroli G, Sorrentino M (2011) Efficiency and evaluation analysis of a network of technology transfer brokers. Technol Anal Strateg Manag 23:7–24

Santoro MD, McGill JP (2005) The effect of uncertainty and asset co-specialization on governance in biotechnology alliances. Strateg Manag J 26:1261–1269

Schouwenburg HC, Groenewoud J (2001) Study motivation under social temptation; effects of trait procrastination. Personality Individ Differ 30:229–240

Seitanidi MM, Crane A (2014) Re-imagining the future of social partnerships and responsible business. Social partnerships and responsible business: a research handbook, pp 388–407

Seitanidi MM, Koufopoulos DN, Palmer P (2010) Partnership formation for change: Indicators for transformative potential in cross sector social partnerships. J Bus Ethics 94:139–161

Seitanidi MM, Lindgreen A (2010) Cross-sector social interactions. J Bus Ethics 94:1–7

Selsky JW, Parker B (2005) Cross-sector partnerships to address social issues: challenges to theory and practice. J Manag 31:849–873

Selsky JW, Parker B (2010) Platforms for cross-sector social partnerships: prospective sensemaking devices for social benefit. J Bus Ethics 94:21–37

Sharfman MP, Gray B, Yan A (1991) The context of interorganizational collaboration in the garment industry: an institutional perspective. J Appl Behav Sci 27:181–208

Siegel DS, Veugelers R, Wright M (2007) Technology transfer offices and commercialization of university intellectual property: performance and policy implications. Oxford Rev Econ Policy 23:640–660

Skelcher C (2005) Public-private partnerships. Oxford University Press, New York, pp 347–370

Sleesman DJ, Conlon DE, McNamara G, Miles JE (2012) Cleaning up the big muddy: a meta-analytic review of the determinants of escalation of commitment. Acad Manag J 55:541–562

Stadtler L, Probst G (2012) How broker organizations can facilitate public–private partnerships for development. Eur Manag J 30:32–46

Staw BM (1976) Knee-deep in the big muddy: a study of escalating commitment to a chosen course of action. Organ Behav Hum Perform 16:27–44

Stone D (2004) Transfer agents and global networks in the 'transnationalization' of policy. J Eur Public Policy 11:545–566

Strongman KT, Burt CD (2000) Taking breaks from work: an exploratory inquiry. J Psychol 134:229–242

Sydow J, Müller-Seitz G, Provan KG (2013) Managing uncertainty in alliances and networks—from governance to practice. Managing knowledge in strategic alliances, 1–43

Ungureanu P, Bertolotti F, Mattarelli E, Bellesia F (2018) Making matters worse by trying to make them better? Exploring vicious circles of decision in hybrid partnerships. Organ Stud. https://doi.org/10.1177/0170840618765575

Van Marrewijk A, Clegg SR, Pitsis TS, Veenswijk M (2008) Managing public–private megaprojects: paradoxes, complexity, and project design. Int J Project Manage 26:591–600

Villani E, Rasmussen E, Grimaldi R (2017) How intermediary organizations facilitate university–industry technology transfer: a proximity approach. Technol Forecast Soc Chang 114:86–102

Vurro C, Dacin MT, Perrini F (2010) Institutional antecedents of partnering for social change: how institutional logics shape cross-sector social partnerships. J Bus Ethics 94:39–53

Waddell S, Brown LD (1997) Fostering intersectoral partnering: a guide to promoting cooperation among government, business, and civil society actors, vol 13. Institute for development research (IDR)

Waddock SA (1989) Understanding social partnerships: an evolutionary model of partnership organizations. Adm Soc 21:78–100

Westley F, Vredenburg H (1991) Strategic bridging: the collaboration between environmentalists and business in the marketing of green products. J Appl Behav Sci 27:65–90

Winch GM (2013) Escalation in major projects: lessons from the Channel Fixed Link. Int J Project Manage 31:724–734

Winer M, Ray K (1994) Collaboration handbook: creating, sustaining, and enjoying the journey. Amherst H. Wilder Foundation, 919 Lafond, St. Paul, MN 55104

York AS, Ahn MJ (2011) University technology transfer office success factors: a comparative case study. Int J Technol Transf Commercialisation 11(1–2):26–50

Part II
Recent Trends in Capacity Building.
The Experience of Progress-TT

CCODE: A New Perspective on Factors that Influence the Growth of Technology Transfer Offices

Alan Kennedy and Peter Frederick

Abstract

CCODE was developed by Pete Frederick and Phil Bullimore of Pera Consulting as a tool to help coaches support small business (SME) growth. It is a robust model of business growth, made up of 20 'factors' that affect the growth potential in SMEs, which are grouped into five categories: Capability Capacity, Opportunity, Desire and Environment. It has been used in PROGRESS-TT to put growth factors identified for Technology Transfer into a context that unified both academic and business aspirations. This chapter describes in detail how CCODE™ was used as a framework for Technology Transfer growth factors and ultimately used to provide the critical areas of focus of the PROGRESS-TT training, coaching and mentoring activities.

Abbreviations

CAF	Critical area of focus
CBS	Capacity building strategy
CBTT	Capacity building technology transfer
EC	European Commission
EIB	European investment bank
GDP	Gross domestic product
IP	Intellectual property
KT	Knowledge transfer
PRO	Public research organisation

A. Kennedy (✉)
MITO Technology, Milan, Italy
e-mail: alan.kennedy@mitotech.eu

P. Frederick
University of Reading, Leicester, UK
e-mail: pete@persuasivewriting.co.uk

© Springer International Publishing AG, part of Springer Nature 2019
M. Granieri and A. Basso (eds.), *Capacity Building in Technology Transfer*,
SxI – Springer for Innovation / SxI – Springer per l'Innovazione 14,
https://doi.org/10.1007/978-3-319-91461-9_5

R&D   Research and development
SME   Small to medium enterprise
TT   Technology transfer
TTNS   Technology transfer national associations
TTO   Technology transfer office

1 Introduction

Europe is spending 0.8% of GDP less than the US and 1.5% less than Japan every year on Research & Development (EC COM2010). The European Union is the largest market in the world but countries like China and South Korea are bringing innovative products to market at a faster rate. The Innovation Union has a target of investing 3% of EU GDP in R&D by 2020 which could create 3.7 million jobs and increase annual GDP by €795 billion by 2025 (EC Innovation Union).

Innovation Union, part of the Europe 2020 plan for smart, sustainable growth, recognises the need to more effectively move IP into industry through Technology Transfer (TT), primarily through contract R&D, licensing and spin-outs. Underpinning that goal is the need for skilled individuals implementing the commercialisation process through Technology Transfer Offices (TTOs) embedded in public research organisations across Europe. Moving more IP into industry requires an increased capacity for Technology transfer.

Technology transfer is the formal and informal movement of know-how, skills, technical knowledge or technology from one organizational setting to another. The process often faces unfavorable economic incentives and an inadequate supply of the resources needed to translate new ideas into technological and economically viable products. The technology transfer process requires access to a number of informational, financial, and human resources and coordination among these various stakeholders can also present a real challenge.

The European Commission launched the CBTT call in Dec 2013 (deadline Mar 2014) to support TTOs in building up their capacity for successful technology transfer. The main objectives of this initiative are to encourage established TTOs to share expertise and best practices with their less experienced counterparts; to boost Europe's ability to turn knowledge into commercial products and services and to prepare Europe's TTOs for InnovFin, the EIBs financing program for innovators.

The PROGRESS-TT initiative—a pilot scheme to develop a growth strategy and implement it through training and mentoring in best practice for successful TT to TTOs and their individual staff members.

Many initiatives intending to better understand the needs of the European TT community or seeking to increase the effectiveness of technology transfer in Europe have taken place over the years. However none of them have specifically aimed at delivering hands-on support for the development of the skills of TT professionals within Public Research Organisations at such as scale as PROGRESS-TT.

PROGRESS-TT implements a strategy derived from the Europe 2020 and Innovation Union framework. The areas of the framework most relevant to PROGRESS-TT are:

Increasing the Capability of stakeholder management

Increasing the Capacity of stakeholders

Increasing the opportunity to Transfer Technology

Increasing the Desire to engage in Technology Transfer.

The goal of the project capacity building programme is to prepare TTOs and TT Funds in public research organisations to attract financing such as InnovFin or other types of early stage investment funds. Capacity building of other stakeholders, like industry or broader finance should be approached through other CBS sub-strategies.

2 Understanding Drivers Behind TTO Growth and Success

The first challenge presented to the PROGRESS-TT partners was to identify the factors that would influence the success of a TTO. The factors affecting growth of businesses have been studied for many years but an equivalent set of tools for TTOs did not exist. As such, the PROGRESS-TT partners decided to start from the premise that the basic principles of growth of an organisation are the same across all sectors and industries, including TTOs. One of the project partners, Pera Consulting, had developed a robust model of business growth, CCODE™. It was decided that this would make a good framework to put the growth factors into a context. Over the past 10 years, CCODE™ has been successfully applied to define the focus and consequent measures of success for growing innovation performance in industry and public sector. CCODE was developed by Pete Frederick and Phil Bullimore of Pera Consulting as a tool to help coaches support small business (SME) growth. It was originally developed as part of the European Commission project Commercialise, a pan-European initiative to provide coaching to recipients of research and development grants. Through extensive research, the Pera Consulting team identified 20 'factors' that affect growth potential in SMEs, which could be grouped into five categories:

1. **Capability** of the management team to manage growth
2. **Capacity** of the business to operate at the desired level
3. **Opportunity** to create growth through exploiting new markets or creating new innovations
4. **Desire** to achieve growth
5. **Environment** where political, physical, and logistical factors are supportive of that growth ambition.

CCODE was also shown to be valid for analysing opportunities to grow, namely Innovation and markets, as well as specific sectors such as Low Carbon, and a range of diagnostic tools have been developed to identify gaps in the growth strategies of SMEs.

The research phase of PROGRESS-TT started with the original CCODE™ model and conducted a detailed literature search to identify which factors in CCODE™ were common to both industry and TTOs. Using this study, they were able to create a tailored version of the tool that better represented the specific factors proven to influence TTO growth (Frederick and Granieri 2015). This so-called TTO CCODE™ is shown below in Table 1.

Table 1 Breakdown and description of CCODE$^{(TM)}$ performance factors

CCODE$^{(TM)}$ Element	Factors to be assessed	Description
Capability of TTO /PRO management	Systems	Management systems and structure for operations and continuous improvement
	Strategy	The existence of and quality of a strategic plan for the TTO
	Skills	Skills of the management team to effectively manage the TTO activities and implement the strategy
	Leadership	The ability of the management to inspire, motivate and drive the TTO staff
Capacity of TTO	Delivery staff numbers	Quantity of personnel
	Marketing and Sales capacity	Availability of marketing and sales functions to generate new business
	Staff Skills	Capability of the staff to do their jobs
	Operations	Existence and quality of operational procedures to implement everyday activities
	Tools and Methodologies	Delivery methodologies able to apply capacity to new opportunities to generate outcomes
	Availability of Funding	Access to finance to implement activities
	Partnerships	Extent of partnership network, giving access to additional resources
Opportunity to transfer technology	Customer Base – Scale	Number of customers for TTO outputs and level of engagement. Includes internal customers (spin-outs)
	Customer base – Absorptive Capacity	Ability of the customers to take up the TTO outputs. Includes internal customers (spin-outs)
	Relationships (networks and understanding)	Scale and quality of relationships with the customer ecosystem, including mutual understanding
	IP Quality	Quality of knowledge produced within the PRO
	Knowledge generation capacity	Quantity of knowledge produced within the PRO
	IP Security	The level of and commitment to protection of IP
	IP Market Relevance	Commercial attractiveness of the knowledge produced
Desire to engage in TT	PRO Management Desire	Willingness of the PRO management to pursue the goals of knowledge transfer
	IP Creator Desire	Willingness of knowledge creators to engage in knowledge transfer
	TTO Desire	Willingness of the TTO management and staff to engage in knowledge transfer
TT Environment	Government support	Government funding/incentives/support programmes to increase knowledge transfer activities
	legislation	Friendliness of the legislative environment to knowledge transfer (e.g. tax breaks, IP regulations)
	Competition	Competitiveness of other knowledge providers and the effect this has on availability of KT opportunities
	Access to services	Presence and quality of supportive services, such as patent offices, export agencies, etc
	Availability of finance	Presence of funds and investors to finance KT activities
	Political/Societal	Economic conditions or political factors affecting TT activities
	Availability of skills	Degree to which the educational system is developing KT skills necessary for the PRO to deliver its long-term aims
	Proximity to market	How close is the PRO to its customers?

3　CCODE™ and PROGRESS-TT

Whilst the TTO CCODE™ provided an excellent starting point, it provided too high a level of detail to be useful as a tool to guide the diagnosis and training process at the heart of PROGRESS-TT. The next step was to use this framework to identify those factors that would be most influential to TTO growth and that could be most easily influenced through intervention.

The partners carried out a second extensive literature search to narrow down the drivers and performance factors that most affect TTO performance, in particular those that work to increase the outputs from commercialisation (formation of spin-out companies, licensing for royalties and direct sale of IP). These were subject to scrutiny by project partners Pera Consulting, Fraunhofer MOEZ and ASTP-Proton to eliminate duplication and to better align them with the needs of TTOs in the framework of the five critical factors of CCODE™. These are presented in Table 2.

This is an extensive list of areas where the performance of a TTO can be influenced for growth but it needed further refinement to consolidate duplicates, combine overlapping factors and remove growth drivers where there is no possibility of providing meaningful intervention or training to the TTO or its parent Public Research Organisation (PRO). For example, there is empirical evidence that the age of the TTO (or the age of the associated PRO) has a positive effect on TT success. However, there is no possibility to change this feature so it does not

Table 2 Factors from literature searches known to affect the performance of TTOs

STAKEHOLDER	CCODE™ Performance Factors for High Growth				
	Capability	Capacity	Opportunity	Desire	Environment
	of stakeholder management	of stakeholders	to transfer technology	to engage in TT	where the PRO and TTO operate
	Factors from literature searches known to affect the performance of TTOs. Therse are framed in the context of CCODE™				
PRO	TT steering tools Age of PRO Size of PRO Entrepreneurial attitude Understanding commercial culture Organization structure Industry relationships	Institutional Prestige	Relationships & Monitoring Institutional Prestige TTO/PRO Website Relationships & Monitoring Disclosure of IP Incentives	Institutional prestige TT steering tools TT as important function Incentives Understanding commercial culture	Geographical Region Research Focus Institutional prestige Establishing industry relationships Monitoring industry relationships
TTO	Organization structure Management Training Benchmarking Skills	Age of TTO Size of TTO Optimal number of staff Marketing Skills Benchmarking Project management TTO Organisation Structure TTO Operational Procedures Incentives	Operational Procedures Employee background Employee Training Stakeholder relationships	Organisational Structure Operational Procedures Focus on licencing and spin-outs Project management skills Benchmarking skills Incentives	TTO/PRO Website Stakeholder relationships Marketing skills
Researcher	TT education	Number of researchers IP producing faculties Entrepreneurial attitude Incentives	Motivation Incentives Number of researchers TT education Entrepreneurial attitude	Entrepreneurial Attitude Motivation	Size of research faculty
Industry	PRO relationships Understanding Academic culture	Stakeholder relationships		Understanding Academic culture Lacking IP expertise Barriers to working with TTO	Collaboration

Table 3 Performance factors filtered and refined

STAKEHOLDER	Performance Factors derived from literature search and filtered to remove duplications and areas where we cannot exert influence				
	Capability	Capacity	Opportunity	Desire	Environment
PRO	Management Training	Research Quality	Create IP	Entrepeneurial Culture	
TTO		Staff Skills		Incentives	
Researcher					Skills Exchange
Industry	IP Management	Capacity to absorb IP		Facilitate IP Exploitation	
Regional / National Networks		Building Networks	Break Down Barriers		Funding for TTO
Private Investor					

translate into a useable growth factor for PROGRESS-TT. At this stage, the range of stakeholders was also expanded to include private networks and regional networks (including local government and funding bodies), leading to a refined set of factors linked to relevant stakeholders, as shown in Table 3.

This concluded the first stage of development of a set of key performance factors affecting TTO growth.

4 From Factors to Tools

The research so far had produced a list of performance factors. The next task was to translate them into topics for training and mentoring and a framework from which Best Practice topics could be derived in the context of increasing the capacity for technology transfer in TTOs. This was greatly facilitated by the consultation phase of the PROGRESS-TT project, where the knowledge derived from research was presented to expert groups for comment and ratification. This was carried out in three different actions:

5 Expert Group Meeting on 18 March 2015

The Expert Group on Defining a Capacity Building Strategy for European TTOs involved 10 top Technology Transfer experts from various EU organisations and 7 representatives from the PROGRESS-TT partner organisations. Together they represented key stakeholders of the TT process across Europe. The meeting was held on 18 March 2015 at the offices of the EC in Brussels. Participants were asked to challenge the proposed ideas and to propose their own solutions to TT effectiveness.

6   Survey of TTO Professionals in Europe

The survey of European TTO professionals was led by the University of Bologna in collaboration with ASTP-Proton and Fraunhofer Moez. The aim of the survey was to collect information on TTOs characteristics, perceived training needs for TTO professionals and current training provision on technology transfer issues for faculty, Ph.Ds and students. The survey was administered between mid-May and early September 2015. Overall, 568 TTOs were contacted and 218 questionnaires collected, 180 of which were fully usable giving a high response rate of 40%.

7   Interviews of Technology Transfer National Associations (TTNAs)

The purpose of these interviews was to gain a comprehensive picture of existing courses on TT-issues, primarily of those offered by the TTNAs themselves, and to identify training needs and therefore any potential skill gaps within TTOs. A total of 23 TTNAs were interviewed with each interview lasting about one hour. This task was undertaken and correlated by University of Barcelona and ASTP Proton over the period June–July 2015.

This consultation process resulted in an understanding of existing best practice and industry thinking. By comparing this to the success factors identified in the previous stage, the project partners were able to create consensus of the perceived gaps in knowledge and training available to TTOs and create a more focused version of the CCODE™ better suited to the actual needs of TTOs across Europe. This, in turn, provided a framework of the critical areas where PROGRESS-TT should focus to identify best practice solutions and to provide topics for training and other support actions. On the advice of all of the experts, 4 of the 5 critical factors of CCODE™ were also modified to better represent the innovation atmosphere and terminology within which TTOs operate: Capability became "TTO Systems" while Capacity and Opportunity were merged to create "Potential". Earlier in the project it became clear that the CCODE™ "Environment" factor was outside of the terms of reference of the PROGRESS-TT project. This factor defines growth drivers which are the remit of governments and wider societal and political bodies. It would not be feasible for PROGRESS-TT to attempt to influence or change them in the limited timescale available to the project. Changing the Environment factor to "Stakeholders" more closely reflected the eco-system in which the TTO works.

The CCODE™ had become the "4 Dimensions":

- TTO Systems
- Potential
- Desire
- Stakeholders

Table 4 4 Dimensions capacity building framework derived from critical growth factors for TTOs

Technology Transfer Cycle	TTO Systems	Potential	Desire	Stakeholders	Focus Area
IDEA ↓		Scouting new ideas and technologies			1
		Incentivising researchers to disclose inventions			
↓	Assessing IP potential				2
↓	Validatng technologies				
↓		Incentivising for commercialisation			
MARKET				Access to Finance	3
				Interactions with investors	
	Securing TTO staff skills				4
	Organising the TTO structure for optimum growth				

Progress-TT now had a framework derived from critical growth factors that could provide focussed actions to promote capacity building in TTOs across Europe. These four dimensions were then cross referenced back to the performance factors defined in the first stage to identify which of these factors would have the greatest effects on the quality and effectiveness of commercialised IP as outputs from TTOs. A consensus was reached through expert panel advice and wide-ranging consultations, resulting in nine performance factors which would provide the basis of PROGRESS-TT actions and Best Practice case studies (see Table 4).

1. Scouting new ideas and technologies
2. Incentivising researchers to disclose inventions
3. Assessing IP potential
4. Validating technologies
5. Incentivising for commercialisation
6. Access to finance
7. Interaction with investors
8. Securing TTO staff skills
9. Organising the TTO structure for optimum growth.

8 PROGRESS-TT Critical Areas of Focus

Results from the PROGRESS-TT Survey to TTO professionals in Europe have shown that, in 2014, over 30% of respondents declared more than 50 contract research agreements completed. This is indicative of a large reliance on Knowledge

Transfer (KT) that provides a one-off up-front income rather than the longer-term licensing income that Technology Transfer can achieve. Outputs from KT are also difficult to measure and do not build reputation for the PRO, an essential performance driver for successful TT. The majority of TTOs in our sample executed less than 3 licensing agreements for royalty payments and created less than 3 spin-off companies using IP derived from their own research faculties.

Concentrating on KT has created a gap in the Technology Transfer value chain and the skills of its practitioners at EU level. The quality of European research outputs is extremely high but the process of taking know-how and IP to industry to make products and services—i.e. commercialisation—needs addressing. The processes involved in TT are as old as commerce itself but the concept of a TTO embedded in a university or other PRO is still a relatively new one in some regions of the EU. Even in PROs where it is considered a strategic mission, there is a lack of capacity to address the needs of TT and facilitate the capture and commercialisation of IP generated by research. The survey supports this as respondents noted the existence of training gaps in the areas of: Marketing technologies, Licensing, and Technology/IPR valuation. The need to offer these courses was rated close to 4 out of 5 (0 being no need and 5 being the greatest perceived need).

The PROGRESS-TT programme has been designed to fill this gap in the value chain and to build the capacity of selected Technology Transfer Offices (TTOs) and TT Funds through intensive support based on a set of best practice cases and dedicated support by selected experts. The best Practice and the support have been offered in areas that have evolved from the critical performance factors of CCODE™ to become the 4 Dimensions of PROGRESS-TT support.

In the original proposal for PROGRESS-TT it was envisaged that candidates for intensive one-to-one coaching and mentoring would first receive a diagnosis of the areas in the TT process where their strengths and weaknesses lay. Support could then be offered in the topics that would benefit the candidate TTO according to their specific needs. TTOs and PRO management are not always immediately receptive to any offer of support that requires them to change. There was a need to market the support on offer from PROGRESS-TTT and the concept of the 4 Dimensions (TT Systems, Potential, Desire, Stakeholders) appeared to offer a simplified approach to the way in which the support was presented. Only two of the nine performance factors selected fitted neatly into these categories. All of the other factors crossed into two or more of the Dimensions. However, when the nine factors were arranged in the order in which they occur in the Technology Transfer Cycle of IDEA → MARKET, with Staff Skills and TTO Organisation supporting the other seven performance factors, obvious dependencies between the factors emerged. There were clearly four Focus Areas where related factors could be grouped. These dependencies and the resulting Focus Areas can be clearly seen in Table 4.

Focus Areas 1,3 and 4 encompass two performance factors each, while Focus Area 2 combines three factors. These were now the PROGRESS-TT "4 Critical Areas of Focus"—the PROGRESS-TT 4 CAFs:

CAF 1: Scouting ideas and technologies and incentivising researchers to disclose IP

CAF 2: Assessing IP potential, validating technologies and incentivising commercialisation

CAF:3 Accessing finance and interacting with financial stakeholders

CAF 4: Securing TTO Staff skills and organising the TTO for optimum growth.

These are 4 critical areas where support for capacity building in technology transfer was either missing or insufficient across the EU and where PROGRESS-TT's efforts should be concentrated to achieve high rates of growth. They are the areas where support for TTOs will have the most impact in building up Europe's capacity for success in technology transfer.

9 TT Diagnostic—Using the 4 CAFs to Determine the Level of Support Needed by TTOs

The PROGRESS-TT support actions included a one-on-one intensive support program for thirty technology transfer organisations and early-stage investment funds. This program offered up to 12 Days coaching and mentoring from a dedicated expert whose skills and experience are matched to the needs of the individual TTO.

There were two calls for an open competition to access this support, one in January 2016 and one in June 2016. Applicants were asked to submit applications via an online selection and diagnostic tool which was based on the PROGRES-TT 4 CAFs. Full details of the TT Diagnostic can be found in Chapter X PROGRESS-TT. Methodology, content, procedures and actions.

References

European Commission (2010) Europe 2020 Flagship Initiative Innovation Union, COM(2010) 546 final

Frederick P, Granieri M (2015) Development of a holistic tool to identify barriers to success for technology transfer offices, final conference 2015 knowledge transfer and entrepreneurship

PROGRESS-TT: Methodology, Content, Procedures, Actions

Andrea Basso, Célia Gavaud and Alan Kennedy

Abstract

In this chapter, we discuss the specifics of the PROGRESS-TT project and in particular the strategy, methodology and the procedures that have been developed for its implementation and how they came to be instrumental to a program to improve the capacity of technology transfer organization.

1 Introduction

Innovation is one of the key elements of Europe's growth strategy and Public Research Organisations (PROs) have been demonstrated in many a study to be an essential generator of innovation (e.g. Allen 1977; Tushman 1977; Tushman and Katz 1980; Jaffe 1989; Adams 1990; Narin et al. 1997; Griliches 1998; Rosenberg and Nelson 1994; Mansfield 1995; Henderson et al. 1998; Branscomb et al. 1999; Cohen et al. 2002; Debackere and Veugelers 2005; Veugelers and Del Rey 2014; European Commission 2014). However, still to this day, Europe is lagging behind

A. Basso (✉) · C. Gavaud · A. Kennedy
MITO Technology, Milan, Italy
e-mail: andrea.basso@mitotech.eu

C. Gavaud
e-mail: celia.gavaud@mitotech.eu

A. Kennedy
e-mail: alan.kennedy@mitotech.eu

© Springer International Publishing AG, part of Springer Nature 2019   83
M. Granieri and A. Basso (eds.), *Capacity Building in Technology Transfer*,
SxI – Springer for Innovation / SxI – Springer per l'Innovazione 14,
https://doi.org/10.1007/978-3-319-91461-9_6

key international players in converting investment in PROs' research results into innovation generating commercial returns.[1]

One way for Europe to turn knowledge into innovation and transfer Intellectual Property (IP) to industry more effectively is by increasing the overall capacity of the technology transfer professionals working in PROs (Volberda et al. 2012; Tom et al. 2012). Upskilling professionals requires leveraging readily available legacy-proven capacity building solutions that would provide them with the appropriate skills, experience, technical training and access to networks (Munari et al. 2016). At the start of PROGRESS-TT in 2014, however, these solutions only existed as a fragmented body of tools and solutions, none of which were embedded into a coherent framework or developed from a strategic, pan-European perspective. On that basis, PROGRESS-TT set out to create a coherent framework of capacity building solutions which would focus on increasing the investment readiness of PROs and their Technology Transfer Offices (TTOs). The project focused on activities aiming at up-skilling TT practitioners in the stages of idea scouting, idea assessing, leading to commercialization and improving access to finance for high-potential TTOs by bridging them with established funds, industry players and investors.

A series of other barriers to technology transfer needed to be overcome in addition to capacity of TT practitioners (Mazurkiewicz et al. 2017). European TTOs are often small and underfunded (Conti and Gaule 2008; Friedman and Silberman 2003). Small and unstructured investment pools and the lack of a common language between IP creators, exploiters and end-users render access to finance difficult (O'Kane et al. 2014; Smilor 2004). In order to address these issues and barriers, the PROGRESS-TT project proposed a new methodology based on best practices and case studies from Europe's most inspiring PROs who manage to overcome these intrinsic and systemic barriers. The best practice and case studies became the basis upon which PROGRESS-TT developed advanced tools, methods and insights that were then tested on several tens of TTOs over the length of the project. Particularly innovative was the project's proposal to tailor audience's readiness level to the support content it delivered. This was achieved by means of a variety of solutions, spanning from traditional training programs to focused workshops and hands-on, immersive bootcamps, all underpinned by an e-learning platform developed ad hoc for the project. In addition, intensive one-to-one support, in the form of coaching and mentoring, was offered to selected high-potential TTOs and TT funds. This would facilitate them to team up with more experienced TTO performers, industry partners and funds to build up their capability, capacity, opportunity and desire for TT and also help them build a more supportive TT ecosystem.

The complete set of solutions have been shaped in a Capacity Building Strategy (see Fig. 1) aiming to pave the way towards creating a coherent and accessible technology transfer capacity building legacy, taking TTOs a step closer to investment such as the InnovFin Technology Transfer Financing Facility (TTFF).

---

[1]European Innovation Scoreboard The so-called 'European paradox': http://ec.europa.eu/growth/industry/innovation/facts-figures/scoreboards_en.

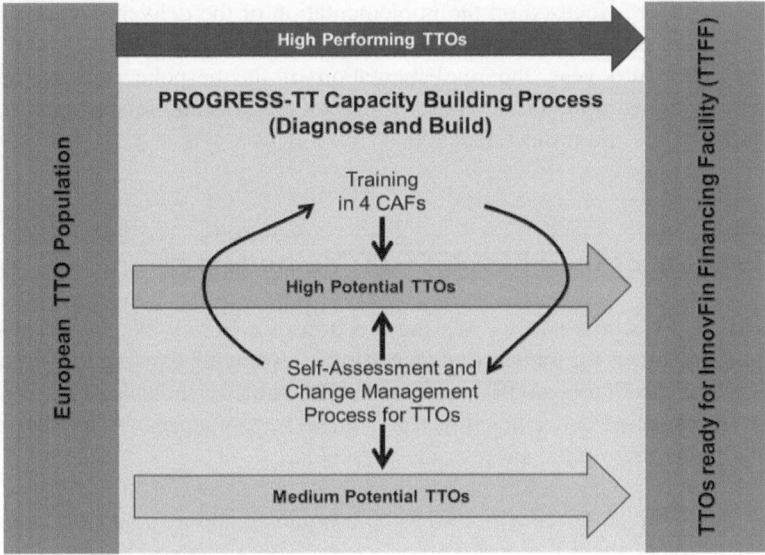

Fig. 1 The PROGRESS-TT capacity building strategy

2 PROGRESS-TT Work Plan

The plan of activities of PROGRESS-TT, presented in Fig. 2, was spread over a 3-year period as follows:

The first year focused mainly on research activities—including the analysis of the prior-art and literature on European TTOs and the identification of the case studies illustrating best practice of the technology transfer process—the results of which contributed to the design of bespoke delivery solutions.

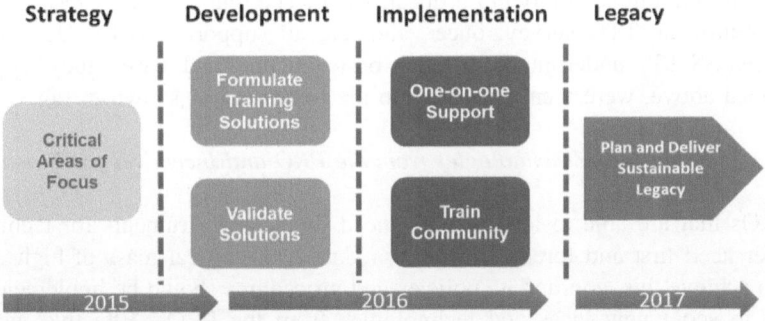

Fig. 2 PROGRESS-TT work plan

The second year focused on the implementation of the delivery solutions based on a 'Diagnose & Build™' holistic approach.

During the third year, the implementation of the bespoke capacity building solutions continued until a final phase of impact analysis and lessons learnt started six months before the project end.

3 Elements of the PROGRESS-TT Methodology

PROGRESS-TT's prime focus was the creation of a library of case studies illustrating best practices for transferring technology, leveraging existing know-how and experience of the European TT community. The strategy shown in Fig. 1 was put together to organize and efficiently share these best practices. A number of key elements underpinning the strategy are presented:

- *The four Critical Areas of Focus*, to help organize and map support against the most stages of the TT process;
- *The Diagnose & Build™* approach to help better fit support to the needs of each individual recipient;
- *The one-to-one coaching & mentoring approach* to offer tailor-made support to high-potential TTOs;
- *The one-to-many capacity building* to ensure wider audiences of TTOs could start on their capacity building journeys.

4 The Four Critical Areas of Focus (CAFs)

Four critical areas of focus (CAFs) were identified at the end of a first year of intensive analysis of the related literature in conjunction with a direct consultation and feedback process with European TTOs as the most relevant steps of the TT process for building capacity in European TTOs. The four areas were tested through a Pan-European TTO survey; once validated, all support activities delivered by PROGRESS-TT, underpinned by the best practice and case study approach described above, were then mapped onto the four areas as shown in Fig. 3.

CAF1: Scouting Ideas/Technologies from the PRO and Incentives for Researchers

PROs that are able to leverage advanced financial instruments for technology transfer need first and foremost to access significant critical mass of high quality IP. To achieve this, appropriate policies and procedures should be implemented by TTOs to scout new ideas and technologies from the PROs. Effective incentive systems should be implemented to engage researchers and faculties in technology

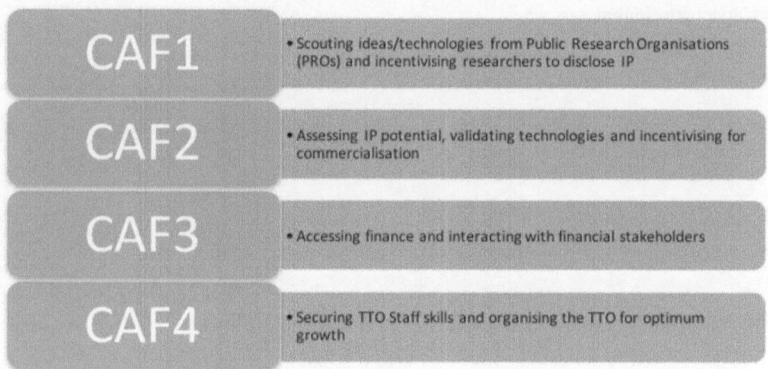

Fig. 3 PROGRESS-TT critical areas of focus (CAFs)

transfer activities. Appropriate incentive schemes are essential for a successful TT value chain from the PRO into financial instruments.

CAF2: Assessing and Validating Technologies and IP

An effective process to assess, test and validate technologies and bring them to a stage where they are market-and investor-ready is critical. This might include proof-of-concept programs, translational activities and other incentives for commercialization.

CAF3: Access to Finance and Interaction with Investors

Technology transfer practitioners need appropriate skills to support academic and student entrepreneurs in interacting, negotiating and closing deals with financial investors, including business angels (BA) venture capital (VC) and investment funds. This should promote the matching process between university spin-offs and start-ups and investors.

CAF4: TTO Staff Skills and Organization of the TTO

Technology transfer teams need to attract, train and retain personnel with specific skills and competences to implement a practice which is appropriate for fostering the commercialization process and working with advanced financial instruments. Planning the governance of the team, its components, its various internal and external interactions, organizing the monitoring and the reward system for staff are all good practice for TT team that more and more must deliver against yearly targets set by the PRO top management.

Fig. 4 Capacity-oriented
Segmentation of TTO
landscape

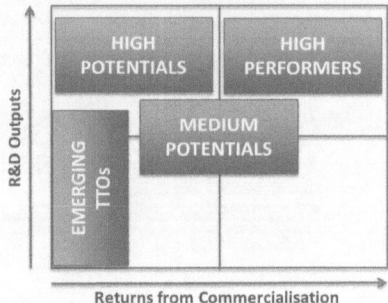

5   The Diagnose & Build™ (D&B) Approach

PROGRESS-TT used a *Diagnose & Build™ approach* which vetted participants to
the project's support programme using online tools.

Assessing TTO's Performance and Performance Potential Using a Diagnostic Tool

The assessment of the TTOs' current performance and performance potential is
the essential first step that will enable TTOs to understand their strengths and
weaknesses and empower them to build capacity in these areas. PROGRESS-TT
created a bespoke diagnostic tool by adapting the well-established CCODE® model
(Frederick and Granieri 2015) for assessment of SME growth factors to PROs and
their TTOs. The PROGRESS-TT diagnostic tool was also used to help the project
select beneficiaries for its intensive one-to-one support programme. The tool was
articulated around two main phases:

TTO Diagnostic Phase 1: Segmenting the European TTO Population

To ensure an optimal use of the limited funding allocated to PROGRESS-TT
(€2.2M over three years), the population of around 700 Public Research Organizations active in Technology Transfer[2] needed to be categorized so available
resources could be concentrated on those TTOs able to swiftly absorb new learning
and practices and move closer to investment readiness.

A segmentation of the TTO population into four different categories, as depicted
in Fig. 4 was implemented with the help of the "TT Diagnostic tool", Phase 1

The TTO Diagnostic Tool Phase 1 segments the European TTO population into
low, medium and high-performers based on scores for four performance factors
used as indicators: Potential, Desire, TTO Systems and Stakeholder engagement—
also known as PROGRESS-TT 4 dimensions. It is implemented as a questionnaire
which relies on a decision tree approach.

---

[2]Source: https://ipib.ci.moez.fraunhofer.de/.

This raw segmentation approach reflects how effectively the TTO can translate the R&D budget of its PRO into commercial success.

The four resulting segments of TTOs are:

High Performing TTOs. High Performers translate the excellent research base of their PRO into high commercial returns. These TTO do not need further support from PROGRESS-TT, but have provided a source of TT best practices for others.

Medium Performing TTOs

High potentials. High Potential TTOs are less effective than the high performers at translating an excellent research base into commercial returns but have the highest potential to become high performers and consequently need intensive and direct support.

Medium potentials. Medium Potential TTOs might have some success in achieving commercial returns, but usually lack a critical mass of high-value IP assets or other know-how with commercial value. A long-term endeavour, with significant engagement of the PRO management can change this situation. However, these TTOs should start a capacity building process as well, but the financial effects will be accountable only at the long term.

Emerging TTOs. TTOs at emerging PROs are in very early stages of their development, usually confronted with a developing research base which has not yet reached a critical mass and with very limited capacities in technology transfer. These TTOs would require extensive support over an extended period. In terms of achieving increased ROI from research expenditures in a reasonable timeframe, the potential of this group is too limited to warrant focusing dedicated project effort in relation to the timeframe of PROGRESS-TT.

On that basis, the targeted capacity building approach of PROGRESS-TT has adopted the following premises:

High performing TTOs will not be supported as they are already capable of attracting and implementing financial instruments.

Due to the limited resources of the project, there will be no or very little support available to Emerging TTOs as they are still too far from the implementation of financial instruments.

Therefore, the PROGRESS-TT strategy for the coaching and mentoring program focuses on the High Potential TTOs, through intensive support, and the Medium Potential TTOs, through wide-reaching support.

Phase 1 assessment allowed to select those TTOs and Funds having a sufficient performance level and performance potential to be taken forward to the stage 2. An example of phase 1 outcome is shown in Fig. 5.

The applicants scoring too low and coming through as emerging TTOs were rejected at this stage. This was a "soft" rejection and the TTOs were supplied with information from the TT Diagnostic indicating where their strengths and weakness were in terms of the four Dimensions (see Fig. 6). They were also advised to visit the PROGRESS-TT website where they could find resources to help them with capacity building activities.

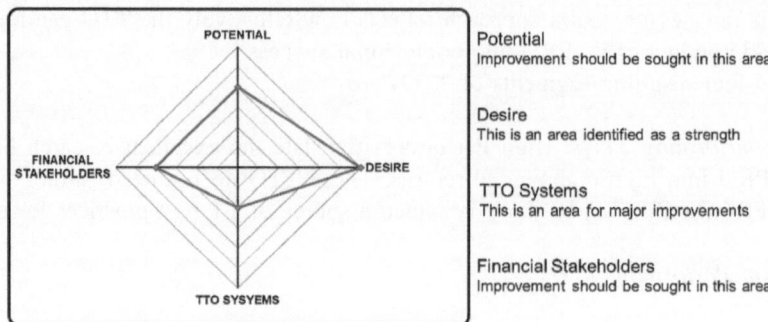

Fig. 5 Phase 1 outputs as seen by applicants

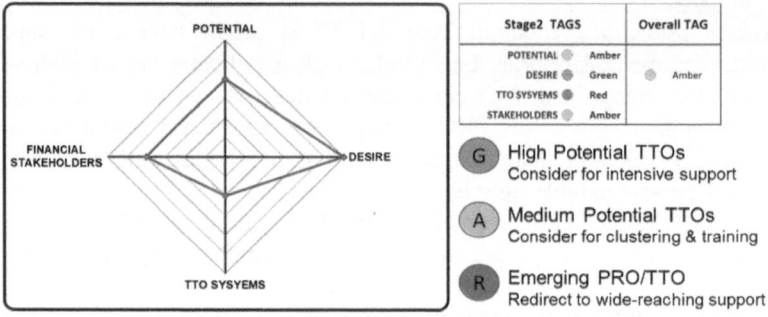

Fig. 6 Phase 1 outputs as seen by project partners

TTO Diagnostic Phase 2: Determining Suitability of TTOs to Receive PROGRESS-TT Support

The TT Diagnostic was also built to support the selection process of the TTOs to receive one-to-one intensive support from PROGRESS-TT. The TT Diagnostic phase 2, in the form of an online questionnaire, determines priority and suitability of candidates to receive PROGRESS-TT intensive support by dividing medium performer TTOs into "high-potential TTOs" and "medium-potential TTOs". All TTOs which passed to Phase 2 of the application process were deemed eligible and suitable for PROGESS-TT mentoring and coaching.

The key elements of phase 2 are shown in Fig. 7 and an example output is shown in Fig. 8. The applicants were asked questions to assess TTO readiness in each of the four Critical Areas of Focus (CAFs). These questions were presented as statements, with a Likert scale against which the applicant rated the performance of their TTO. The guidelines supplied for the application process stressed that these answers were made as subjectively as possible.

4 Critical Areas of Focus

1	Scouting ideas and technologies
	Incentivising researchers to disclose IP
2	Assessing IP potential and validating technologies
	Incentivising for commercialisation
3	Accessing Finance
	Interacting with financial stakeholders
4	Securing TTO staff skills
	Organising the TTO for optimum growth

G High Potentials
Candidate for intensive support

A Medium Potentials
Candidate for clustering

R Emerging PRO/TTO
Signpost to wide-ranging support

Fig. 7 Phase 2 process

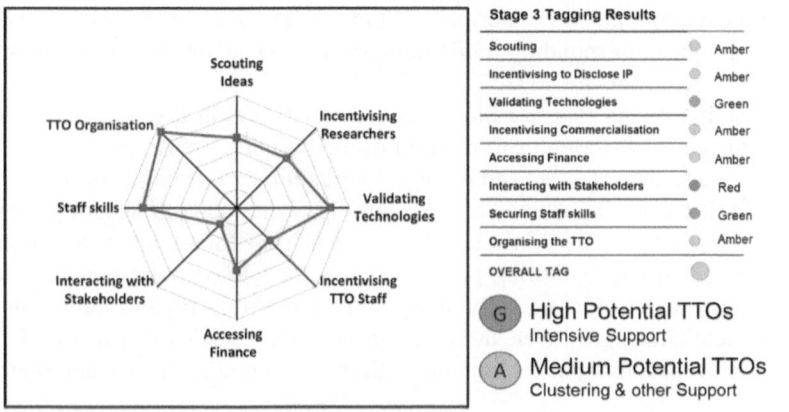

Fig. 8 Phase 2 output example as seen by project partners

The completed applications were reviewed manually by a panel of experts. The use of an expert panel augments the automatic online process and produces a finer review of the TTOs. For example, In CAF1 the questions about the TTO capacity for Scouting New Technologies were as shown in Table 1.

Table 1 Phase 2 filter—questions on scouting new technologies

Critical Area of Focus #1:	Please put a number in each box on the right corresponding to the level of Staffing resources and systems available to the TTO				ENTER (1 - 4)
A. Scouting Ideas from Public Research Organisations	1	2	3	4	
The TTO has sufficient staff focused on scouting new ideas and inventions from faculty	None	Adequate	Good	Excellent	
The TTO has the resources to follow up on the disclosures generated by scouting	None	Adequate	Good	Excellent	
The TTO has systems in place to manage invention disclosures and their associated IP	None	Adequate	Good	Excellent	

Table 2 Score values and corresponding traffic light colours

Normalised Score	Traffic Light		Comment
0.76 – 1.00	G	Area of strength	TTO has excellent expertise and staff skills
0.26 – 0.75	A	Area for improvement	TTO has strong skills but needs to increase capacity
0.00 – 0.25	R	Area of weakness	TTO needs a lot of work to increase capacity

The output of the Phase 2 filter initially used a radar-like report as shown in Figs. 6 and 8 that was eventually changed to a "Traffic Light" system. Scores in each of the eight topics were normalized to the range 0–1. Table 2 shows the score values with the corresponding traffic light and comments on the implications of the traffic light score for the TTO.

The output of the Phase 2 filter was presented to both the applicant TTO and their assigned Expert/Mentor to be used as the basis for development of an action plan to address the areas in need of capacity building. An example of this output is shown in Table 3. This shows a TTO with excellent skills in scouting ideas and incentivizing researchers to disclose IP, suggesting that these areas do not need to be addressed in the mentoring program. They have strong skills in the areas of assessing IP and validating technology and have strong organizational and staff development skills. However, these are areas where capacity can readily be built. The areas of incentivizing for commercialization, accessing finance and interacting

Table 3 Phase 2 filter traffic light output example

Phase 2 Traffic Lights

CAF	Description	Score	
1A	Scouting ideas and technologies	G	0.75
1B	Incentivising researchers to disclose IP	G	0.78
2A	Assessing IP potential and validating technologies	A	0.50
2B	Incentivising for commercialisation	R	0.45
3A	Accessing Finance	R	0.30
3B	Interacting with financial stakeholders	R	0.40
4A	Securing TTO staff skills	A	0.70
4B	Organising the TTO for optimum growth	A	0.68

with financial stakeholders are poorly developed and will need a lot of work to develop capacity. In creating an action plan to address these needs, the expert and the TTO would be encouraged to concentrate on CAF2A, CAF4A and CAF4B.

Validating the TTO Diagnostic Approach

The scoring system has been validated on a test group of TTOs which completed the application process and then was invited to discuss the outputs and identify possible improvements. Three TTOs were recruited at test phase to give feedback on the tool structure and content. Test TTOs reviews were analysed and changes made to the tools, where necessary; the three test TTOs were later re-engaged to fill in data in the tools. Results were then compared with the known profiles of the TTOs and adjustments made to the scoring system. All the TTOs in the test group commented that they were surprised by some of the tool outcomes but also said that, on reflection, it gave a fair representation of their actual status.

The TT diagnostic tool was designed to be fit-for-purpose within the available resources of the PROGRESS-TT action. During the project, the output of this innovative but limited model was validated by the intervention of a panel of experts who reviewed the results manually. As part of the project exploitation, after the end of the project, the TT Diagnostic could be further developed into a more sophisticated, cross-validated model. Figures 9, 10 and 11 are screenshots of the user interface of the tool.

In the following sections we review the vetting procedure used for the one-to-many type of events (e.g. workshops, training, and bootcamps).

D&B: Vetting Event Participants

All candidate participants were required, before being able to register for a PROGRESS-TT event, to complete the Event Participant Vetting tool to ensure that the best solutions were offered to TT practitioners. Importantly, the vetting tool helped assess the needs of the TTOs as a whole and not those of the individual staff members attending the events. The results of the vetting tool were then used to adapt the content of the events to the needs of the participant TTOs. The vetting tool primarily looked at specific TTO strategy and mandates. TTOs in Europe have quite diverse mandates which can be categorized into three main areas of TTO activity— Research Support, Revenue Generation and Social and Economic Benefit (Giuri et al. 2016). There is often a poor understanding on how these activities relate together in building a TTO and PRO growth strategy. On the other side the correct determination of the current and desired future ratio of these activities is a crucial factor for any capacity building strategy in technology transfer to be effective and sustainable. A better understanding of the mandates of TTOs applying to register to an event helped create "immersive experiences" at the bootcamps—where participants had similar profiles in terms of expertise and needs—and more homogeneous groups at workshops and training sessions.

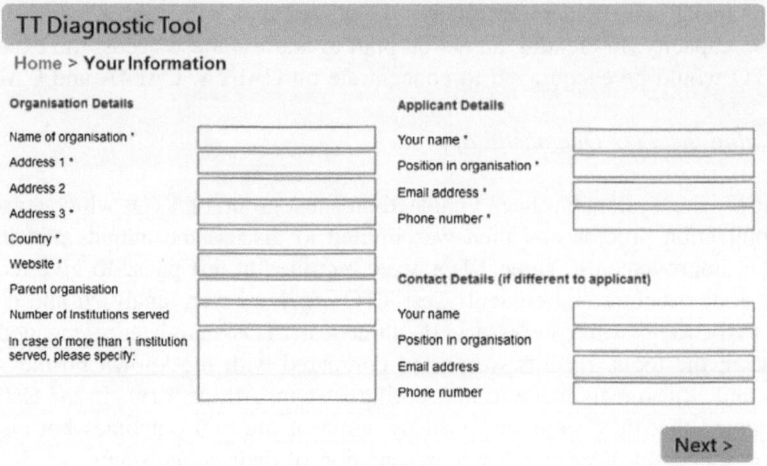

Fig. 9 Example user interface of TT Diagnostic tool

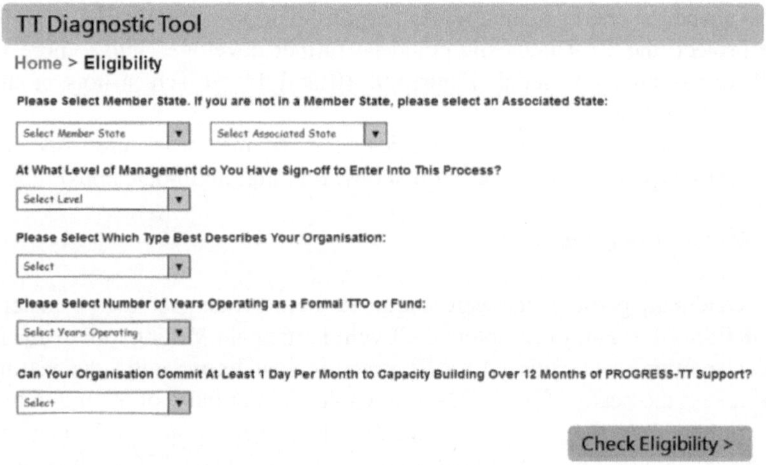

Fig. 10 Another example of user interface of TT Diagnostic tool

Customized Journeys

A side outcome of the Diagnose & Build™ approach was the emergence of customized 'journeys' for growth for TTOs. The original intention of D&B was to provide support for TTOs tailored to their individual needs. Time is a scarce resource in a busy TTO and this was envisaged as a tool to attract TTO senior staff away from their day job to participate to training events. The effect was more

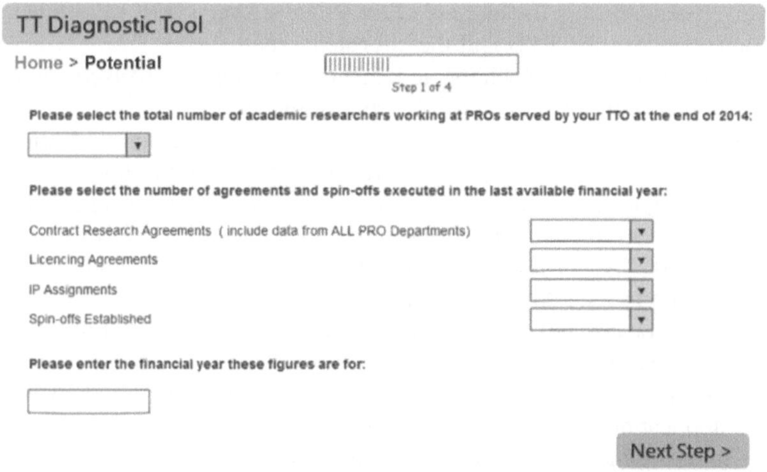

Fig. 11 Example 3 user interface of TT Diagnostic tool

profound than that, as it encouraged some TTOs to take part in many PROGRESS-TT events. It became apparent that a few TTOs were serial participants in various project events, for example, sending staff to consecutive bootcamps as well as participating in the coaching and mentoring programmes. This "customized journey" approach was helping them to build capacity in a more holistic manner. This was not included as part of the original PROGRESS-TT proposal but spontaneously emerged from the actual delivery of the programme. It worked very well for the TTOs involved and it is our belief that it should be further explored in any future capacity building actions.

6 PROGRESS-TT One-to-One Support: Coaching and Mentoring of Individual TTOs

An innovative and essential part of the PROGRESS-TT capacity building strategy focused on a programme of intensive one-on-one coaching and mentoring of technology transfer teams and early-stage investment funds. This programme offered up to 12 days of coaching and mentoring from a dedicated expert whose skills and experience were matched to the needs of individual TTOs. There were two calls for an open competition to access this support, one in January 2016 and one in June 2016. Interested TTOs submitted their applications via an online portal. Following an automatic and panel selection, thirty high-potential performers were invited to receive PROGRESS-TT mentoring and coaching support with priority in one of the four critical areas of focus (CAFs). Experts with specialised knowledge were individually matched to participating TTOs and support activities were aligned with the needs of the beneficiaries and with individual capacity building roadmaps generated from the outputs of the Phase 2 Diagnostic and refined during discussions between the TTO and the allocated expert.

The methodology for coaching and mentoring support revolved around two models; the Teaming model and the Clustering model.

The Teaming model was offered to single PROs or funds to work on specific weaknesses identified. The assigned expert organized a first meeting with the TTO to refine the diagnostic phase and agree an action plan with the recipient PRO. The expert then built a Team of stakeholders around the recipient PRO to address the issues identified. The team usually consisted of another PRO (high-performing), and "peer mentors" including an industrial mentor (Multinational Corporation) and a financial mentor (Funds). It could also include a spin-off or start-up company if the main exploitation model chosen by the TTO is through a company. The teaming model answered a recurrent need of PROs to enhance their ties between basic research, application-orientated research and industry.

The Clustering model was offered to groups of PROs sharing the issue of lack of critical mass of exploitable IP. Based on a proven model of IP pooling, an expert in facilitating that model acted as facilitator for the group of PROs and worked with them on establishing a roadmap to clustering. On-site activities were complemented with on-line support.

7   PROGRESS TT One-to-Many Support: Training, Workshops and Bootcamps

7.1   PROGRESS-TT Live Training Events

PROGRESS-TT live training events were led by seasoned TT practitioners with long training facilitation experience. Sessions focused on a given Critical Area of Focus, providing professionals with tools and knowledge that they can then incorporate into their work and processes for immediate use. Participation was restricted to 30 participants for optimum experience. Training events were targeted at professionals working in TTO with between one to three years of experience in the field. Live training events dealt with situations participants face in their day-to-day work, focusing on the strategic skills and tools required by professionals in the TTO.

Live Training Format

Live training sessions combined a lecture from an expert in a given field with practical exercises. The use of PROGRESS-TT best practice case studies as material throughout the day enabled the application of theoretical concepts in real-life scenarios. The roles and responsibilities of the key players during a workshop were as follows:

Expert: Prepare and deliver content, use training tools & methods, share relevant experience;

Technical Support (offsite and onsite): preparation and logistics of the event; during the event itself, the focus was on resolving issues (possibly using work-arounds).

Live Training Approach

Each individual training session focused on one of the PROGRESS-TT CAFs covering idea scouting, assessment or financing. Training sessions aimed to provide attendees with a practical plan of implementation combined with tools and expert knowledge that could be easily incorporated into their work processes for impact on the TTOs. Live training not only covered market oriented theoretical and practical knowledge but sought to make participants aware that they were not alone. The events brought together other participants facing similar realities and problems in other regions across Europe.

As with all PROGRESS-TT events, a case study discussion approach allowed the application of theoretical concepts to be demonstrated in real-life cases to bridge the gap between theory and practice and encourage active learning. It also provided an opportunity for the development of key skills such as communication, group working and problem solving, and hopefully increased the trainees' enjoyment of the session and their desire to learn from the experience. Experts were asked to structure the session over 6 h (not including lunch break) as follows:

Morning Session

- 3.5 h theory presentation (including 30 min for networking/coffee break):

  - Explanation of the main issues at stake (concepts, main problems to overcome, strategic aspects of the area);
  - Presentation of selected case studies and tools commonly used to face issues presented and how to work with the tools;
  - Opening of discussion with the participants on how they can use the case studies for inspiration and tools in their institutions;
  - Wrap-up and summary of the main concepts.

Afternoon Session

- The expert was asked to present in detail at least three PROGRESS-TT case studies and corresponding Best Practice from the online Library[3] (1 h per Case Study) and to explain his/her take on the actions taken by the TTOs involved in the cases (e.g. highlight the background to the case, the key success factors and limitations, etc.);

---

[3]Available at http://www.progresstt.eu/best-practice-library/.

Discussions on replicability were encouraged.

Throughout the day, the expert was expected to ensure some level of interaction (for example to allow questions during the presentation of the theoretical concepts and during the cases discussion).

7.2 PROGRESS-TT Workshops

PROGRESS-TT workshops were highly interactive and high-impact one-day sessions, led by a moderator and an expert, targeted at professionals working in TTOs with at least 5 years of experience in the field. Participation was restricted to 15 participants for optimum experience.

Workshop Approach

Each workshop was based on one of the PROGRESS-TT CAFs covering idea scouting, assessment or financing. Similar to the Live Training sessions, workshops were aimed at providing attendees with a practical plan of implementation combined with tools and expert knowledge that could be easily incorporated into their work processes for impact on the TTOs. The events brought them together with other participants facing similar realities and problems in other regions across Europe.

As with all PROGRESS-TT events, a case study discussion approach allowed the application of theoretical concepts to be demonstrated in real-life cases. This helped bridge the gap between theory and practice and encouraging active learning. It also provided an opportunity for the development of key skills such as communication, group working and problem solving, and hopefully increased the trainees' enjoyment of the session and their desire to learn from the experience.

Workshop Format

PROGRESS-TT workshops were live, physical event shaped as highly interactive sessions combining the sharing of experience and knowledge between the participants with practical exercises and theoretical content given by a selected moderator. The participatory approach helped create a favourable environment to brainstorming and supportive of exchanges of ideas, problems and solutions to facilitate TTOs decision-making processes. The roles and responsibilities of the key players during a workshop were as follows:

- Moderator: Usually a PROGRESS-TT partner (hosting the event), well-aware of the case studies to be presented; the moderator managed the introduction and closeout, helped ensure that expert and participants were interacting appropriately and summarised the questions for a Q&A session;
- Expert: Prepared and delivered event content, using PROGRESS-TT training tools and methods (e.g. case studies), sharing relevant experience;

- Technical Support: for the preparation and logistics of the event; during the event itself the focus was on resolving issues (possibly using work-arounds).

Expert and moderator were asked to structure the session over 6 h (not including lunch break) as follows:

Morning Session

- A theory-based presentation introduced the core issues discussed in the workshop, the state of the art of the CAF, an analysis of the case studies selected by the expert and the main conclusions that can be drawn from the cases to be discussed in the second part of the morning.

Afternoon Session

- Participants were asked to split into groups to solve real-life problems based on PROGRESS-TT case studies and the direct experience of participant TTOs.
- A roundtable discussion was then held to come up with possible solutions (e.g. to incentivise the disclosure of IP). Participants were asked to divide into groups to solve the cases. A responsible for each group presented the conclusions in a final PowerPoint format in the last hour.
- Expert and moderator were there at all time to assist in the discussions and provide feedback on the outputs at the end of each discussion. They ensured the session was as lively and inclusive as possible.

7.3 PROGRESS-TT Bootcamps

PROGRESS-TT three-day immersive bootcamps were targeted primarily at TTO managers and senior staff able to take the learning from the bootcamps back to their own organisations. The goal was sustainable building of Technology Transfer capacity in PROs, rather than personal development of individual TT executives. The target was to ensure that all attendees interacted with other participants, expert speakers and representatives of the host institutions to create a "Deep Dive" experience into all aspects of the TT process.

During the project, four bootcamps were held at the premises of private sector partners with extensive track record in technology transfer: the first was held in June 2016 at DSM Nutritional Products in Basel, Switzerland, a second bootcamp was held in November 2016 at Philips in Eindhoven, The Netherlands; a third was hosted in April 2017 by VTT ventures in Espoo, Finland and a final event was hosted by Philips in September 2017 in Eindhoven, the Netherlands.

Bootcamp Format

Each of the three bootcamp days concentrated on one of the first three Critical Areas of Focus, i.e. covering either scouting, assessment or finance. The fourth CAF, dealing with the horizontal issue of staff skills and team organisation, was embedded in each of the other three. The morning presentations and interactive discussions with TT experts were followed by afternoon workshops promoting the development of tools and organizational skills related to the day's topic.

Day 1 started with introductions; Day 2 and Day 3 with a brief re-cap of the previous day's learning. Each day was run by a bootcamp facilitator who ensured continuity of the whole bootcamp and who made sure that discussions stayed on theme and kept to the allocated time slots. In addition, each day an external expert was asked to deliver the first morning session on a topic related to the CAF at stake. An example agenda of the morning sessions is shown below in Table 4.

Expert sessions were structured as follows:

- 1.5 h presentation including 20 min for discussion;
- Discussion leveraging three PROGRESS-TT case studies and their corresponding Best Practice from the online library.

Following each morning session, participants were asked to divide into groups to solve real-life problems. These were based on PROGRESS-TT case studies and the direct experience of the problems faced by European TTOs. The expert assisted in the facilitation and provided feedback on the outputs at the end of the day. An example of bootcamp topics is presented in Table 5.

The afternoon sessions began with a presentation on the day's topic by an expert from the hosting partner and continued with a workshop on the development of tools which can be used to streamline TT activities discussed during the day's sessions. Examples of the afternoon workshop topics in Table 6 below.

Evening Networking Sessions

Evening networking sessions were organized around 6 pm on Day 1 and Day 2 after the participants had a chance to freshen up and catch-up on emails and work. The evening sessions, greatly appreciated by participants, helped debrief the learning of the day and continue exploring solutions to problems faced by TTOs in a more relaxed atmosphere.

Table 4 Morning session topics

DAY	CAF	Morning session topic	Date	Expert
Day 1	CAF 1	Filling the funnel with new projects	8 NOV	Jeff Skinner
Day 2	CAF 2	Choosing the right projects to take forward and the best route to commercialisation	9 NOV	Bart Nelissen
Day 3	CAF 3	Understanding what industry wants to secure investment	10 NOV	Jane Davies

Table 5 Bootcamp topics

DAY	CAF	Morning Bootcamp Topic	Expert
Day 1	CAF 1	Incentivizing researchers to disclose: your solutions!" (based on PROGRESS-TT case studies)	Jeff Skinner
Day 2	CAF 2	Your good practice for assessing technology"; how do they fit PROGRESS-TT and translate for use by other TTOs?	Bart Nelissen
Day 3	CAF 3	Use Best Practice Actions, your own experience and insights gained at the Bootcamp to build an Action Plan to implement change in your TTO for accessing finance and liaising with investors	Jane Davies

Table 6 Afternoon bootcamp topics

DAY	CAF	Afternoon bootcamp topic (tools)
Day 1	CAF 1	Creating and using tools to gather information and to streamline the initial assessment of new disclosures
Day 2	CAF 2	Creating and using tools to improve due diligence in technology and IP assessment
Day 3	CAF 3	Using tools to prepare for investment

8 Conclusions and Lessons Learned

At the end of the implementation phase (end of 2017) of the capacity building activities a full analysis of the impact of PROGRESS-TT actions will be carried out and recommendations for further and potentially refined support will be made by PROGRESS-TT. However several lessons can be drawn from the PROGRESS-TT experience:

Identification of Skills Gaps and Matching Training to Individual TTOs

There are essentially two parts to the Diagnose and Build (D&B) approach as implemented by PROGRESS-TT. The first part focused on the identification of areas of higher risks in the TT process or areas of the TT process that have the largest skills gaps at European level. The second part of the D&B strategy is the actual diagnostic process where the TTO population was segmented into High Performers, High Potentials, Medium Potentials and Emerging TTOs. The community of TTOs is so disparate that a specific categorisation, going beyond considerations of structure, governance, activities and organisation, is essential to normalise assessment of teams of TT practitioners. Once the segmentation process was complete, TTOs could easily be matched with the identified areas of higher risk and larger skills gaps which then enabled customisation of training packages to suit each TTOs specific needs.

Packaging suites of solutions and making them available singly or in a journey type approach has provided the modularity which TTOs require to learn whilst going about their daily job. As an effective capacity building solution, the coaching has been particularly appreciated and is seen by the participant TTOs to contribute greatly to achieving higher performance in their TT activities. (This observation comes as direct feedback from coached TTOs and is currently being evidenced through various analyses. This will be completed before the end of the project in December 2017). The "Culture Gap" between TTOs and the "real world" is huge but is being bridged through newer generation of TTO staff. Capacity building is needed throughout the 4 CAF to ensure all-rounded TTO staff who can drive their TTOs to higher performance and efficiency levels.

Understanding the Nature and Status of a TTO is Crucial to the Delivery of Effective Capacity Building Solutions

The mandate that a TTO has from the PRO senior management is often badly articulated or has not been addressed at all. Some activities, commonly carried out by TTOs in Europe, such as Research Support and miscellaneous "Non-TT Tasks", can use up a huge amount of time and resources on activities which do not in any way contribute to feeding the funnel of ideas for TT. In extreme cases, this could be mundane duties such as conducting tours of the University and Innovation centres for visitors. More often theses are tasks that, although important to the running of the organisation, don't fit easily into other departments and are "dumped" on the TTO. During delivery of workshops, bootcamps and mentoring services in PROGRESS-TT, we have learned that many TTOs lack strategic aims in the formation of these mandates. This leaves the TTO vulnerable to non-constructive change and using up scarce resources on tasks not related to TT. Highlighting the ratios of these activities to PRO management has proved to be a useful tool in re-structuring mandates for individual TTOs to promote increasing capacity for TT. Direct involvement with the PRO management on a sustainable basis is needed to ensure that strategy focusses on the capacity to be successful in TT.

It is a crucial factor for growth that the TTO mandate is regarded as a strategic mission in the overall activities of the PRO and that it works continuously to affirm and sustain its credibility and visibility to PRO management and researchers.

Deal-Readiness is a Remote Perspective for Most European TTOs

The gap between research outputs and investment-ready solutions is universal in the TTO process. Research outputs tend to be at Technology Readiness Level (TRL) 5, even after the validation and proof of concept (PoC) provided by many PROs/TTOs. Industry expects a TRL 7 and up and ready to drop into their processes or portfolio of products with minimal further development.

Insights from PROGRESS-TT research show that training courses on TT issues for faculty, PhD and students mostly covered IPR protection, start-up creation and industry consulting/research contracts. However, access to finance and interactions with investors is only marginally covered by such courses. Early networking with investors, business leaders and other financial stakeholders is a crucial part of this cycle but TTOs often lack the entrepreneurial attitude needed to setup and take advantage of this kind of networking. TTOs often neglect networking with their local innovation eco-system including PRO Management, National Associations, local government, innovation agencies, European technology platforms and the entrepreneurial environment of business and funding agencies. TTOs need to embed methodologies and tools to interact with the ecosystem to market their projects and attract expertise and finance for commercialization. The local ecosystem, in terms of industries and innovation, should play a key role in the success of the TTO. A weak ecosystem that is not able to foster innovation and is not able to absorb the outcomes of university research impacts negatively the success of the TTO. On the supply side of the ecosystem innovation/IPR generation requires a good quality research that motivates industry to be part of the technology transfer process.

A Holistic Approach to Building Capacity in the TTO Requires a Different set of Solutions to those Needed for Building Skills in Individual TT Professionals

Although we don't have hard evidence of this (PROGRESS-TT was not designed to gather this information), the experience of many of the PROGRESS-TT experts involved in the delivery of the coaching and mentoring and the bootcamps has been that TTOs are often lacking a clearly defined, end-to end TT process and workflow. The individual TT professionals may be highly skilled but the TTO lacks a systematic approach to many of the tasks which makes up the TT process of "ideas to market". PROGRESS-TT, with the framework based on the four Critical Areas of Focus has provided guidelines on how to build and strengthen such workflow. Delivery solutions such as bootcamps provide an immersive experience into the whole process of TT, from idea disclosure to deal making and placing products on the market. This has highlighted the culture gap to the participants (who are TTO managers and senior staff) who have reported that the bootcamps helped them dramatically in closing this culture gap and in networking to join up all their activities to create a streamlined and more effective commercialization process.

References

Adams J (1990) Fundamental stocks of knowledge and productivity growth. J Polit Econ 98:673–702
Allen T (1977) Managing the flow of technology. MIT Press
Branscomb LM, Kodama F, Florida R (1999) Industrializing knowledge. The MIT Press, Cambridge
Cohen W, Nelson R, Walsh J (2002) Links and impacts: the influence of public research on industrial R&D. Manage Sci 48(1):1–23
Conti A, Gaule P (2008) The CEMI survey of university technology transfer offices in Europe, Report CEMI report, CEMI_TTO2008
Debackere K, Veugelers R (2005) The role of academic technology transfer organizations in improving industry science links. Res Policy 34:321–342

European Commission (2014) Research and innovation as sources of renewed growth. Report, European Commission, SWD(2014)181, Brussels 2014 SWD(2014)181

Frederick P, Granieri M (2015) Development of a holistic tool to identify barriers to success for technology transfer offices. Social Science Research Network, SSRN, April 2015 https://papers.ssrn.com/sol3/papers.cfm?abstract_id=2596508

Friedman J, Silberman J (2003) University technology transfer: do incentives, management, and location matter? J Technol Transf 28(1):17–30

Giuri P, Munari F, Toschi L (2016) The strategic orientation of universities in knowledge transfer activities SPRU 50th anniversary conference. Brigton, UK, 6–9 Sept 2016

Griliches Z (1998) R&D and productivity. Chicago University Press

Henderson R, Jaffe AB, Trajtenberg M (1998) Universities as a source of commercial technology: a detailed analysis of university patenting, 1965–1988. Rev Econ Stat 80(1):119–127

Jaffe AB (1989) Real effects of academic research. Am Econ Rev 79(5):957–970

Mansfield E (1995) Academic research underlying industrial innovations: sources, characteristics, and financing. Rev Econ Stat 55–56

Mazurkiewicz A, Poteralska B (2017) Technology transfer barriers and challenges faced by R&D organisations. In: Procedia engineering, vol 182, pp 457–465. ISSN:1877-7058, http://dx.doi.org/10.1016/j.proeng.2017.03.134, http://www.sciencedirect.com/science/article/pii/S1877705817312705

Munari F, Rasmussen E, Toschi L, Villani E (2016) Determinants of the university technology transfer policy-mix: a cross-national analysis of gap-funding instruments. J Technol Transf 41 (6):1377–1405. http://EconPapers.repec.org/RePEc:kap:jtecht:v:41:y:2016:i:6:d:10.1007_s10961-015-9448-1

Narin F, Hamilton KS, Olivastro D (1997) The increasing linkage between US technology and science. Res Policy 26:317–330

O'Kane C, Mangematin V, Geoghegan W, Fitzgerald C (2014) University technology transfer offices: the search for identity to build legitimacy. Res Policy. https://doi.org/10.1016/j.respol.2014.08.003OKane2014

Rosenberg N, Nelson R (1994) American universities and technical advance in industry. Res Policy 23:323–348

Smilor R (2004) University venturing: technology transfer and commercialization in higher education. Int J Technol Transf Commercial 3(1):111

Tom JMM, Oshri I, Volberda HW (2012) The skills base of technology transfer professionals. pp 871–891 | Published online: 08 Oct 2012

Tushman M (1977) Special boundary roles in the innovation process. Adm Sci Q 22:587–605

Tushman M, Katz R (1980) External communication and project performance: an investigation into the role of gatekeepers. Manage Sci 26(11):1071–1085

Volberda HK, Oshri I, Mom JM (2012) Technology transfer: the practice and the profession. pp 863–869 | Published online: 08 Oct 2012

Veugelers R, Del Rey (2014) The contribution of universities to innovation, (regional) growth and employment. European Networks on Economics of Education University of Girona (2014)

The Intersection Between Capacity Building and Finance

Federico Munari and Laura Toschi

Abstract

New ventures do not have access to the same financial resources as larger firms, for the presence of the so-called "funding gap". Such a barrier can be particularly pronounced for university startups, given their knowledge- and technology-based nature. To address this market failure, universities (often in collaboration with public and/or private partners) have started to activate programs to support their startups. By presenting a set of best practices and case studies, the chapter describes three main areas of intervention in the area of capacity building for accessing finance: (i) raising awareness and competences to access external finance (i.e. training activities and commercialization boot-camps), (ii) supporting the validation and maturation of university technologies (i.e. proof-of-concept programmes and university accelerator programmes) and (iii) enhancing interactions and building partnership with investors (i.e. matchmaking events with investors, networks of different types of investors, formalized partnerships between universities and financial investors, and university seed funds).

F. Munari · L. Toschi (✉)
Department of Management, University of Bologna, Bologna, Italy
e-mail: laura.toschi@unibo.it

F. Munari
e-mail: federico.munari@unibo.it

© Springer International Publishing AG, part of Springer Nature 2019   105
M. Granieri and A. Basso (eds.), *Capacity Building in Technology Transfer*,
SxI – Springer for Innovation / SxI – Springer per l'Innovazione 14,
https://doi.org/10.1007/978-3-319-91461-9_7

1   Introduction

To start and grow an innovative business requires financial capital. Access to finance relates to the ability of young and innovative ventures to obtain the most appropriate source of finance applicable to their stage of development. New ventures, by their nature, do not have access to the same financial resources as larger or more mature firms. Indeed, for new companies operating in knowledge-based and high-tech sectors, a series of barriers dealing with high uncertainty levels, presence of information asymmetries and lack of successful track records or warranties, limit the ability to obtain funding from traditional sources. The presence of a so-called "funding gap" can be particularly pronounced in the case of university spinoffs and startups. Technologies and startups stemming from universities, indeed, are typically characterized by uncertainty and informational gaps, which make it difficult for external investors to assess the business prospects or monitor the entrepreneurs once the investments are made (Lerner 2005; Munari and Toschi 2011). The existence of these barriers and market failures has led many universities, often in collaboration with public and private partners, to activate specific programs offering financial support and assistance in order to enhance the investment readiness of university startups and technologies and facilitate their ability to access external finance.

In this Chapter, we will analyse and discuss the main areas of intervention of universities and related intermediation structures—such as technology transfer offices (TTOs), incubators, entrepreneurship centres—in the area of capacity building for accessing finance. We will focus on answering the following key questions: "What kind of solutions should be implemented by universities and TTOs to effectively support academics and students in successfully interacting and closing deals with investors? What are the key elements needed to achieve the effective matching between university spinoffs/startups and financial investors? How to prepare teams of university researchers and inventors to be investor-ready?". To answer such questions, we present a set of best practices in this domain, implemented by European universities and TTOs and emerging from the analyses conducted within the Progress-TT project (see Chapter "Selected Case Studies" by Marcello Torrisi for a description of the process for the identification of the best practices and related case-studies). In the rest of this Chapter we first discuss the origins of the funding gap for university technologies, and we then present the set of university initiatives at the intersection of capacity building and access to finance. In this respect, we follow a conceptual model centred on three core objectives and areas of intervention: Raising awareness and competences to access external finance; Supporting the validation and maturation of university technologies; Enhancing interactions and building partnership with investors.

2 The Origins of the Funding Gap for University Technologies

A new venture typically follows a life cycle characterized by different stages of development. Depending on the stage in which the new venture is proceeding, different types of investors should be more suitable to address its funding needs (Berger and Udell 1998). It is, thus, possible to distinguish between the stages of pre-seed, seed, growth and exit. In the more embryonic phases (i.e. pre-seed), in which the company itself is still in the design phase and has not yet been incorporated and the entrepreneurial project has just a patented idea or an invention, without a final product that can be commercialized, personal funds, friend and family funding are the most suitable forms of financing. In the more recent years, crowd-funding has also become an interesting opportunity to fund entrepreneurial projects in this initial phase. In the seed phase, the technical, commercial and economic feasibility of entrepreneurial ideas is tested and early revenues may start to be generated from the commercialization on new products or services. Business angels (i.e. individuals, such as former entrepreneurs or professionals, providing starting equity financing) typically play a critical role in this phase, together with funding from public grants supporting entrepreneurship and innovation. Later on, when the company more steadily grows, generating more significant revenue streams, venture capitalists and other institutional investors become more active actors in the funding process. Finally, in the maturity stage, the company could more easily access traditional forms of funding, such as bank funding. In this phase, more successful and rapidly growing companies may also consider the possibility to access public markets through an Initial Public Offer for fueling their growth (see Fig. 1). Along

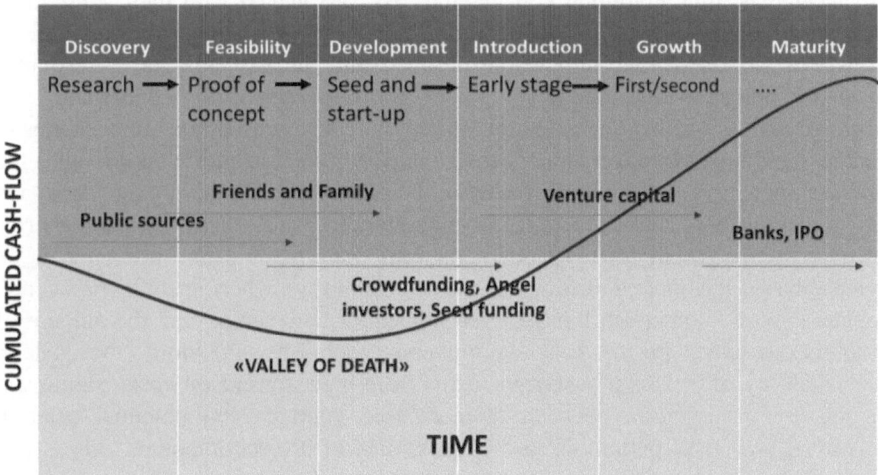

Fig. 1 Different types of investors during the lifecycle of a new venture (adapted from www. ibusinessangel.com)

this life cycle, however, very often emerges a funding gap, which may significantly slow-down or even interrupt the growth process of new ventures. The funding gap emerges when "*the funding requirements of a company are greater than those that can be met by the small scale providers of finance, but not substantially enough to be considered by the large equity providers*" (Bank of England 2000, p. iii).

The motives underlying the difficulty experienced in accessing external risk capital can be distinguished between factors that affect the supply of finance and factors that affect the demand for finance. Factors from the *supply* side that lead traditional investors to prefer larger deals in order to finance later-stage companies rather than smaller deals for early-stage and innovative companies are high risks, high information asymmetries (both as adverse selection and moral hazard), high transaction costs and ongoing running costs, and lack of exit options. From the *demand* side, instead, the main factor that reduces the attractiveness of newly established, high-tech ventures to formal equity financing, is related to the lack of investment readiness by these companies, like a limited understanding of equity instruments, reluctance to cede control or ownership, lack of business and commercial competences and experience, poor quality of businesses and poor communication skills.

In sum, external investors, including those more advanced and risk-oriented such as venture capitalists or business angels, may be reluctant to invest in very early-stage and science-based technologies. Due to the presence of this well-known trap (often labelled also as the "Valley of Death" for early-stage technologies) inventions and technologies that may later be socially and economically useful, but are not yet commercially viable, may stall.

Such problems, that are common for all type of high-tech startups, can be magnified in the case of university spinoffs and startups generated from frontier research projects. In fact, such companies tend to be at the frontier of scientific advancements, thus involving considerable risks associated with their subsequent validation, industrialization, and commercialization. The time lag required to transform the discoveries into viable products and the vast amount of resources needed to pursue the required development, generate a mix of high uncertainty and negative cash flows that decreases investment incentives and limits opportunities to secure funding. Moreover, the strong emphasis on scientific skills and the non-commercial nature of academic spinoffs' expose the founding teams to a lack of competencies to make the business plan attractive and clear to financing firms (Munari and Toschi 2011). Such patterns are particularly pronounced in science-based sectors (e.g., life sciences, biotechnology), where specific market and regulatory conditions push the bar even higher for both timing and the amount of resources needed. In the case of university spin-offs, therefore, the general unavailability of private investments stems from high transaction costs, significant asymmetric information between science-based ventures and potential external investors, high risks pertaining to the uncertainty of project outcomes.

To address this challenges, various universities and public research organizations (PROs) have implemented support programs and created internal financial mechanisms (i.e., "gap funding" instruments) in order to strengthen the investor-readiness of academic spin-offs, and startups and enhance their capacity to access external finance, often in collaboration with public institutions (Darcy et al. 2009; Lerner 2009; Wright et al. 2006; Munari et al. 2016). At the light of the abovementioned considerations, it is clear that such support programs should address not only the limited availability of external funding, but also the limited quality of investment opportunities. Providing more cash alone will not be sufficient to solve the funding gap, given that many businesses fail to raise external equity finance because they are not ready to be financed. Thus, entrepreneurs need to be educated as to the nature and needs of potential investors. The "state of mind" of the academic entrepreneur toward access to finance should be prepared in order to address three main typical entrepreneurial failings: (i) lack of awareness and entrepreneur's attitude towards equity finance, which refers to the limited knowledge and understanding of different forms of financing, (ii) presentational failings, which include both shortcomings business plans and deficiencies in oral presentations and (iii) 'investability' of the project, which refers to the entrepreneurial team and business ability to meets the requirements of external investors (Mason and Harrison 2001).

An effective intervention designed by universities and/or public institutions to enhance all the three areas identified in the case of university spinoffs and startups should thus follow a multi-stage model for capacity building in access to finance (see Fig. 2), consisting of different activities to address the entrepreneurs' gaps toward financing. These activities are linked together in order to determine a process along which entrepreneurial teams and companies can proceed to increase their preparation toward investments. The first stage is centred on the delivery of dedicated trainings (i.e. seminars, workshops, bootcamps) to enhance the general awareness of university would-be entrepreneurs (including researchers and students) about equity financing and their level of skills and competences in the area of commercialization. Once this step has been finalized, entrepreneurial projects should be evaluated and filtered to understand their growth potential and suitability to access additional funding (and in particular equity funding). Ideas and technologies reaching a positive assessment can proceed to the subsequent phase, where

Area of intervention	Raising awareness and competences	Supporting the validation and maturation of ideas and technologies	Enhancing interactions and building partnerships with investors
Main objectives	• Increase awareness • Enhance skills and competences • Improve presentational skills	• Test and validate ideas • Define valorisation plan • Increase investor readiness	• Facilitate intermediation • Establish direct links with investors • Increase availability of seed funding
Programs and initiatives	• Seminars and courses • Commercialisation bootcamps	• Proof-of-concept funding • Accelerator programs	• Matchmaking events • Investors' networks • Partnerships with investors • University Seed Funds

Fig. 2 A conceptual model for capacity building in access to finance

ad hoc programs—such as proof-of-concept or accelerator programs—can support their validation and maturation. Such programs are also critical to assist academic entrepreneurs to define their commercialisation plans and prepare winning investment presentations, in order to signal personal and organizational competences to investors. Following such preparatory stages, a next step should favour the creation of direct linkages between entrepreneurial teams and financial investors, by allowing businesses ready to be financed to meet potential investors and concretely capture their attention. The finalization of investment opportunities in this phase could also be favoured by universities through the creation of partnerships with financial stakeholders and, in some selected cases, through the co-investment in seed funding instruments.

Following this conceptual model, in the next sections of the Chapter, we will analyse in more depth the three areas of intervention for universities in the area of capacity building for accessing finance:

- Raising awareness and competences to access external finance
- Supporting the validation and maturation of ideas and technologies
- Enhancing interactions and building partnership with investors.

2.1 Raising Awareness and Competences to Access External Finance

A first and important set of initiatives can be activated by universities and public research centres in order to increase the awareness and skills to support academic and student entrepreneurs in the domain of access to finance. As previously described, the existence of a funding gap is not exclusively due to a limited availability of sources of financing, but also to a limited availability of valuable opportunities of investment. In other words, investors are often unable to invest at the frequency that they would like to, due to the lack of quality opportunities that they see in the market (Mason and Harrison 2001). The goal, in this case, is, thus, to fill the gap between research and investors and increase the pool of investable businesses. **Training activities** may act as an important tool to implement, with the aim to increase awareness and knowledge on business and financial issues among the academic staff members and the students. They include a set of training programmes and initiatives (seminars, courses, webinars, bootcamps, etc.), dealing also with topics related to access to finance, aimed at raising the quality of investment opportunities generated from the academia. The goal is to prepare entrepreneurs of university spinoffs and start-ups for the investment stage, therefore raising their skill and competence levels, and by that their investment readiness. They aim at addressing some of the criticalities that limit the ability of university spinoffs/start-up to access external (equity) finance:

- Limited awareness of funding sources and equity aversion: a first issue concerns the lack of awareness and sufficient knowledge by researches and students of the different available funding options, and of which of them better suit their needs. An additional issue relates to equity aversion phenomenon. While all entrepreneurs (both academic and not) have aspirations to grow their businesses, most of them are reluctant to surrender ownership and control. Equity aversion, also depends from the entrepreneur's lack of information about the characteristics and availability of alternative sources of finance. The consequence is that many potentially investable projects do not come forward as potential recipients for business angels and venture capital funding.
- Lack of competences and limited investment readiness: from the investor perspective, most businesses that seek external finance do not meet the requirements of external investors, as shown by the high rejection rates of business angels and venture capital funds. Among the reasons for investor reluctance in investing, the composition of the team has been revealed as particularly critical. Very often, academic entrepreneurial teams tend to be strong in terms of technical competences, but lack adequate business, commercial and financial competences.
- Presentation and communication failings: even if the underlying proposition is sound, a business may still fail to raise finance if the business plan is poorly constructed and presented. Such communication problems include shortcomings in business plans and other written documents that are aimed at investors and also deficiencies in pitches at investment forums.

In order to deal with such issue, therefore, it is important that universities and related intermediation structures (TTOs, incubators, entrepreneurship centres) organize dedicated training activities including also dedicated sessions devoted to access to finance and interactions with financial investors. The central objective of such training programmes is, thus, to raise the skills and competences of entrepreneurial teams, and by that the quality of investment opportunities. University TTOs, as intermediary in the technology transfer process, can play a critical role also in organizing and/or supporting this type of training programs. The study by Bolzani et al. (2017) on a sample of 176 university TTOs in Europe shows that the provision of training by TTOs is a key factor in promoting the impact from science-based entrepreneurship. However, access to finance is one of the least covered topics in this type of training activities, that most frequently focus on issues related to IPR protection and start-up formation. The results, based on a survey on TTO managers in Europe, highlight that only about 30% of the responding TTOs cover topics related to access to finance in the training activities in which they are involved.

A training programme to enhance investment readiness should involve at least two elements. The first is information provision: entrepreneurs do not know about the advantages of equity finance vis-à-vis other types of funding provision, what is required to attract equity investors, the criteria that investors use to assess investment opportunities nor how to sell their investment proposal to investors. The second is support: helping entrepreneurs to meet these standards. Accordingly, investment readiness programmes are developed in order to offer several capacity building activities, including seminars and training on funding opportunities, support for effective business planning and pitching and dedicated mentoring and coaching. An example in this sense is provided by TechMinho, the TTO of the University of Minho in Portugal, organising a broad range of training activities to support entrepreneurs-to-be and improve the ability of academics to communicate with investors: (i) workshops and seminars where serial entrepreneurs share their knowledge on how to build a business, (ii) a business plan competition, (iii) an Idea Lab for early projects, a sort of lean acceleration programme taking place once a week over four months where academics are matched with mentors and coaches to intensely work on their business models, (iv) a Company Lab for more advanced spin-off projects focusing on practical aspects of running a business, (v) a helpdesk offering individual consulting services on specific issues relating to company foundation and growth.

Among the different set of training and support activities, **Commercialisation Bootcamps** are particularly relevant. They are intensive and short (i.e. from two-days to five-days, often spread throughout a period of several weeks) training and development programmes, which introduce researchers and students to the various stages of the commercialisation process, dealing with topics such as protecting IP, structuring a licensing agreement or a start-up team, realising the potential of commercial outcomes in the market, interacting with financial stakeholders. At the end of the boot camp, participants should have developed a solid awareness of what it takes to build a commercialisation plan and have matured a clearer perspective on the potential and feasibility of the idea. During these intensive programmes, participants are typically required to apply the principles learnt to their own situation in interactive, practical and hands-on workshop sessions, in order to enhance their commercialisation capability. In addition, the program typically ends with a pitch session, where participants have the opportunity to present their projects to a panel of investors.

An example is the bootcamp launched by UCDInnovation, the TTO of the University College Dublin (UCD). The UCD Commercialisation Bootcamp is a support program for academics and staff at University College Dublin and National College of Art and Design (NCAD) with the aim to provide their researchers, staff and post graduate research students with the knowledge, skills and understanding of the technology commercialisation process and strengthen the pipeline of commercial opportunities arising from their research programmes. It takes place twice a year and it consists of five 3-h practical and hands-on workshops delivered over a 5-week period, where the 10 project teams selected by year are mentored in-between by UCD Innovation case managers. The areas in which commercialization awareness is developed are mainly related to the identification of the market needs and market validation, technology roadmap, commercial exploitation route(s), project team skills and funding plan. The idea is to develop a commercial plan to obtain an early and fast feedback from potential users and investors on their ideas, so to identify promising areas for development and abandon less interesting avenues.

2.2 Supporting the Maturation of University Technologies: The Role of Proof of Concept and Accelerator Programs

A second important group of activities and tools for addressing the funding gap aims at supporting the initial assessment of the commercialisation potential of ideas and technologies from PROs and favouring their technical and commercial maturation. Activities related to the assessment, testing, validation and maturation of technologies are fundamental to help overcoming the specific hurdles to the transfer of innovations outside of the University. In their early-stages, university-generated inventions are generally still too embryonic in nature and often at the frontier of scientific advancements to earn money through commercial sales or attract the attention of private external investors, as shown through the concept of funding gap. As a consequence, the provision of dedicated maturation activities and resources by the TTO (often in collaboration with the University and with other external organisations or public agencies), to help university inventions mature to a stage where they are market and investor ready, represents a major facilitator for effective technology transfer. The availability of well-designed and well-functioning technology validation and maturation programmes therefore represents a critical component in a technology transfer ecosystem.

Two main initiatives can be identified in this area, representing different types of technology validation and maturation mechanisms used (individually or simultaneously) by TTOs and Universities to demonstrate an invention's technical and market viability: Proof-of-concept programmes and University Accelerator Programmes. **Proof-of-concept (PoC) programmes (POCs)** represent innovative mechanisms that encompass several funding schemes that combine money, expertise, and training to help new inventions and discoveries emerge and demonstrate their technical and commercial feasibility (Munari et al. 2016, 2017a, b). Despite diverse labels across different universities and innovation agencies (e.g., proof-of-concept funds, proof-of-principle funds, translational funding, pre-seed funding, verification funding, maturation programmes, innovation grants, ignition grants), they share common objectives and characteristics. These programmes aim at evaluating the technical feasibility and the commercial potential of early-stage university/PRO ideas and technologies, so to demonstrate their value to potential industrial partners and investors. Such programmes, thus, provide capital and assistance to individual researchers or research teams across a whole spectrum of areas, such as IP protection, prototype building and technical verification, business plan development, market studies, entrepreneurial team formation and networking with external partners. The ultimate goal is to bring the technology up to a point where it can be licensed to external industrial partners or where a start-up can be created, so to attract the interest of investors in the following stages of development. A successful example of PoC is the VFT-1 (Verifiering för Tillväxt) Programme, established in 2012 by the KTH Royal Institute of Technology, with the support of the Swedish Innovation Agency Vinnova. The goal of the VFT-1 is to take a project to commercialisation or prepare it for a more rigorous commercial and technical verification. Typically, a project remains within the VFT-1 for about 1–1.5 years receiving initial small financing rounds that last 2–3 months, followed by larger financing rounds that last 5–9 months. The VFT-1 funds about 40–50 projects per year with an average investment per project of around 12,000 EUR. Other examples are offered by University College London and EPFL. University College London has established different types of internal funding schemes in order to support the transition of early-stage technologies from their labs to the marketplace. The maximum levels of POC grants managed by UCL are in the range going from £25,000 to £100,000, depending on the type of program. Similarly, through the Innogrant scheme, established in 2005 with the support of Lombard Odier, EPFL financed more than 80 teams and helped to create more than 50 start-ups. An Innogrant is provided up to a maximum of CHF100,000 (around 92,000 Euro), mostly in the form of an EPFL salary for young researchers working on valorisation activities.

There is a wide variety of configurations of PoC programmes across universities and countries, along a series of dimensions: managing organisation; type of funding; sources of funding; amount of funding; program's focus and eligible projects; selection criteria; selection committee; additional support activities and project management. This variety is often determined by regulations and expectations of public funding bodies backing the programmes. In terms of organisational arrangements, PoC programmes can be operated and administered locally by university TTOs or centrally by regional or national entities. PoC programmes may also be operated locally by the TTOs or (in most of the cases) be partially or totally funded by external public or private organisations. Looking at sources of capital for the establishment of university-managed proof-of-concept programmes, generally they largely come from the Universities and from public institutions, at either the national or regional level (Munari et al. 2016). Sometimes, there is an involvement of private actors. The evidence suggests that this type of funding is an investment area where the involvement of public sector sources is of utmost importance, due to high uncertainty levels, long-term horizons, and high potential impact for social welfare.

PoC funding programmes may also differ in terms of project eligibility. Some programmes provide PoC funding only for projects developed by university researchers, other programmes support projects stemming from anyone within the university, including undergraduate and graduate students. Generally, PoC programmes for students and for researchers differ substantially in the amount of funding. Also, POC funds are typically administered in the form of grants, although various typologies are available, such as repayment schemes or soft loans. As far as the selection of projects for PoC programmes is concerned, some TTOs launch calls for applications, followed by a selection process among submitted projects. Other TTOs prefer to select projects for PoC funding on a rolling basis, from a pool of projects that are being developed with the support of the TTO. Regardless of the selection process adopted, a common feature of these programmes is the adoption of a clear and transparent pool of criteria and the use of customized tools and scales to assess their technological and commercial readiness levels. Moreover, the selection of POC projects can be undertaken directly by the TTO or involve a broader committee composed also by scientists and external technology transfer experts (e.g. consultants, industry representatives, business angels or venture capitalists).

Finally, in order to assure an appropriate use of PoC funding, the funding distribution process should be tightly managed and controlled by the TTO. There are risks in providing large amounts of money directly to researchers to be used at their discretion, since money could be spent for further basic research work rather than for technology transfer activities. Stage funding provided upon the receipt of well-defined milestones, instead, is a good practice that allows to control for the risks faced by the project and check whether the PoC funding is appropriately spent.

A similar initiative is represented by **University Accelerator programmes**, supporting the creation and launch of sustainable and profitable new ventures based on university research and intellectual property. They are fixed-term (generally, 3–4 months), cohort-based programmes that include mentorship and educational components, and end with a public pitch event or demo day. The objective of these programmes is to equip researchers with the knowledge, skills and understanding that will be required to work as part of a team leading a new commercial venture. Typically, projects within an accelerator programmes receive also training and expert mentoring on issues related to marketing, finance and funding, sales and pitching to investors. Such programmes may also provide funding to the entrepreneurial teams selected into the program, either as an initial investment to all selected teams or as a prize awarded to the winner of a final competition. Moreover, accelerator programmes may favour the construction of linkages between the new ventures and industry partners/venture capitalists. During accelerator programmes, researchers get to cooperate with industry partners or serial entrepreneurs, who can help them further develop their business ideas. Two additional specific value-added characteristics of university accelerators are (i) the possibility to involve as mentors distinguished alumni entrepreneurs, managers or investors and (ii) the possibility for start-up's teams to have access to students' talent and physical infrastructure such as labs and workshops.

An example is the TechFounders Program at Technical University Munich (TUM). The TechFounders programme is a 3-month international accelerator programme in Munich which brings tech start-ups, industry and venture capitalists together. For the TechFounders Program, start-ups with a fit to one of the industry partners of UnternehmerTUM (the centre for innovation and business creation at TUM) and in the early stage phases are eligible. Selected participants get 25.000 Euro funding and access to office space and the possibility to attend the high-tech workshops organized by UnternehmerTUM and to obtain on-demand expert coaching for maturing their technologies. In the programme, the start-ups will work closely together with partners from the industry with the aim to cooperate with one of them and attract the interest of a high-class customer or investor.

2.3   Enhancing Interactions and Building Partnerships with Investors

In order to facilitate the access to financial sources for science-based projects and startups, universities and related intermediation structures (TTOs, incubators, entrepreneurship centres) may activate an additional range of initiatives and programmes to favour the establishment of direct and fruitful interactions between researchers and investors. These can range from the organization of dedicated matchmaking events to the establishment of formalized partnerships. The aim of these initiatives is to help researchers to establish fruitful connections with the community of investors and, therefore, to move quickly early-stage spinout projects and research ideas to a position where they could attract interest from external investors.

The simplest and less demanding activity in this area is represented by the organisation of **matchmaking events with investors**, where researchers and students present their ideas (pitches) to potential investors. Matchmaking events often consist of general pitching sessions and more dedicated "speed-dating" sessions, where potential entrepreneurs (e.g. academics, researchers) have the chance to meet and interact with several potential investors to present their own ideas and, eventually, ask for their support. It is a way to match demand and supply of financial resources to support early-stage spin-off, spinouts and start-up ventures coming from academia. For the matching to be successful, it is fundamental that ventures are in a position where they could attract investors. For this reason, TTOs may offer dedicated pitch and communication trainings to prepare the academics/ entrepreneurs for the event. An example is represented by TecMinho, that has established in 2010 a yearly half-day event called "Investors' Day" to connect with investors academics who have early-stage validated technologies available for investment. During the "Investors' Day", 20–30 investors, ranging from business angels and venture capitalists to decision makers representing public funding agencies or micro-credit opportunities, will actively participate to the pitches presented and indicate which projects they would like to meet immediately to further analyse the selected projects. Before the event, the TTO offers dedicated pitch and communication trainings to prepare the academics/entrepreneurs for the Day.

A more formalised measure regards the establishment of **networks of business angels, venture capitalists and other investors**. Formalised networks are a useful resource to tap into every time there is the need to receive additional information on funding academic spin-offs and start-ups, to propose research ideas, and to proceed with the creation of spinoffs and start-ups. An example is represented by Oxford Angels Network, involving investors and private companies interested in investing in spin-out companies from the University of Oxford. Business angels participating in the network are entitled to receive business proposals circulated by Oxford University Innovation (the Technology Transfer company of the University of Oxford) on behalf of spinout companies looking for funding. In addition to that, they can receive invitations to regular Investment Meetings, at which both early-stage and established spinout companies present their funding opportunities.

A further step is represented by the establishment of **formalized partnerships between PROs and VC firms and other financial investors**. These partnership schemes aim at establishing and nurturing fruitful and systematic linkages and collaborations with investors, with the final goal to enhance the opportunities to finance all promising technologies and spinout projects. Throughout these schemes, TTOs/PROs strengthen their network of potential investors, thereby being able to increase the pipeline of successful projects carried out. Partnerships can be of two types: "loose partnerships" or "exclusive partnership". In the former case, regular connections are established with VC firms and investors, although with limited degree of formalization and commitment by both partners. Typically, these are not exclusive partnerships, meaning that there is no obligation for investors to steadily support academic research ideas, and there is limited or no financial commitment by universities. However, they provide the chance for PROs/TTOs to propose spin-off and start-up ideas to a pool of investors without having to search for potentially interested actors from scratch. For instance, High-Tech Gründerfonds (HTGF) is a public-private venture capital investment company founded in 2005 and based in Germany, has established special agreements with ten German institutions - including RTWH Aachen University, Free University of Berlin, Technical University of Munich, Fraunhofer Venture and Max-Planck-Innovation. In such special partnerships, the typical interlocutor for HTGF is represented by the University TTO, the incubator or the Centre for Entrepreneurship. The cornerstone of the partnership is represented by the organisation of regular matching events (such as one-day long pitching events,) organised by the institutions and attended by representatives of HTGF as a way to scout promising investment opportunities.

On the other hand, "exclusive partnerships" are based on long-term commitments with VC firms, where investors may also acquire the right to commercialise, through the creation of spin-out companies, the PRO's research-generated intellectual property. This type of partnership is based on strong commitment by both sides of the investment process, thus potentially boosting TT activities more generally. An example is represented by the ten-year agreement between Cardiff University and Fusion IP (now part of IP Group, a publicly-listed intellectual property company specialized in the commercialisation of IP sourced from research institutions). It was completed in 2007 and consists in a ten-year exclusive agreement with Fusion IP (which was acquired successively by IP Group), giving it the right to commercialise, through the creation of spin-out companies, Cardiff University's research-generated intellectual property. According to the terms of this agreement, Fusion IP had the first right to establish a spinout company based on Cardiff University owned intellectual property. In return for this exclusive option, Fusion IP gave the University a significant shareholding in the Fund. Furthermore, it provided a ring-fenced funding for investing in spinout commercialisation opportunities arising from Cardiff University's research

The final step in the process of partnership creation with investors, involving a higher commitment (also in financial terms) by the PRO, is represented by the creation of **University Seed Funds (USFs)**. University seed funds are gap funding equity instruments organised or sponsored by PROs/TTOs to fund university spinoffs and start-ups, and contrast the lack of private funding in support of the transition of early stage university technology from the lab to the marketplace (Munari et al. 2015; Croce et al. 2014). The paper by Munari et al. (2015) has identified 73 USFs in European countries since 1990, showing a wide variety of institutional configurations. Some USFs can be directly managed by a TT unit linked to the PRO (as in the case of funds managed by Imperial Innovation at Imperial College London). Here the university typically provides part of the initial capital injection into the fund, although in most of the cases this is complemented by other sources of funding coming from (most frequently) public sources or from private and other sources. Most of USFs are established as externally managed funds. In such cases, the funds are managed by a specialized investor group and the PRO is typically involved as a limited partner. An example of such configuration is represented by 3T Capital, a French venture capital fund serving innovative companies within the ICT sector from their seed and early-stage phases. 3T Capital is funded by the European Investment Fund, BPI France, and its strategic university partner, Institut Mines Telecom.

The results of the study by Munari et al. (2015) on USFs in Europe highlight useful lessons for the design and implementation of such funding instruments by PROs. First, the size and quality of the research base upon which the USF is built, which emerges as another important precondition for delivering effective results by means of this type of financial instrument. The availability of high-quality science at the university or in the subject region is thus an important precondition for the success of such financial instruments, guaranteeing a steady stream of high-potential companies and the possibility of developing a diversified portfolio of high-quality companies in which to invest. Second, USFs should be adequately funded to support the projected lifetime, the expectations of the stakeholders involved and the future potential for sustainability. Reaching a critical mass for USFs is important for two main sets of reasons. First, a larger size of the fund allows to diversify the investments among different companies and to provide at the same time adequate financial resources to investees. Second, it allows attracting and involving professionalized VC investors to manage the fund.

3 Conclusions

In this Chapter, we have analyzed a comprehensive range of solutions which universities and TTOs can implement to support university-based projects in their challenge to capture external investors' attention and address the well-known funding gap. The approach applied in the Chapter is based on the analysis of relevant best practices gathered across European academic contexts according with a multi-stage conceptual model centred on three main areas of intervention. The first area aims to help academics raising awareness and competences to access external finance; the second is focused on supporting the validation and maturation of university technologies; the third is driven by the need to enhance interactions and build partnership with investors. From this comprehensive approach, we can derive some important conclusions.

First, access to finance continues to be one of the most relevant problems faced by academic spinoffs in their attempt to commercialize their technologies. Although the supply side of the problem (i.e. increasing the availability of financial resources) continues to be relevant, we show that the demand side has to get more and more attention. Capacity building in this area is, thus, a critical area of intervention for universities in the domain of Third Mission activities. Universities devoting strategic importance to entrepreneurship and knowledge transfer should increasingly invest resources for capacity building. However, universities need to be aware that in order to obtain results in this field a gradual process is needed. The three areas of interventions proposed in the chapter can be included in the university's agenda in subsequent steps and with a certain level of independence among them according with the requests of each university.

Second, and related to the previous point, there exists a variety of programs and initiatives that can be implemented in this area, addressing different aspects of the funding gap. This panel of possibilities highlights the importance to acknowledge the existence not only of a funding gap, but also a knowledge and communication gap. Given this heterogeneity of the problem, a systemic approach should be pursued (probably through gradual steps), as suggested by our conceptual framework.

Third, it is also valuable to recognize the importance to align the mix of initiatives for capacity building with the characteristics of the universities: size, level and type of specialization, research quality, size and expertise of TTOs and other intermediation structures. In particular, a clear attention should be devoted to the university strategic orientation in the area of technology transfer. Some universities have detailed strategic plans in this area, while others integrate the Third Mission in their agenda in a more general nuance. Depending on the academic context under consideration, it is thus critical to adapt the capacity building framework according with the final goals of the university.

Forth, the characteristics of the external entrepreneurial ecosystem also matter. In particular, it is important to involve as much as possible various actors from the ecosystem in the initiatives for capacity building in access to finance, starting from financial investors, but including also public stakeholders, alumni, successful entrepreneurs, business partners. Also in this case, depending on the specific area of intervention, a different type of stakeholder can have a more or less relevant impact and significantly contribute to the resolution of the funding gap.

Finally, it is important to recognize that the results achievable through the different activities proposed are not likely to manifest in the short-term, so vision and continuity of effort are required not only in the implementation phase but also in the final phase of harvesting of the benefits. Between the two steps, however, a systematic approach to monitor and control the outcomes and final impact of such activities, through a dedicated set of indicators and metrics, has to be considered in order to be sure to proceed in the right direction.

References

Australian Venture Capital Association Limited. http://www.avcal.com.au/

Bank of England (2000) Finance for small firms: a seventh report. Bank of England, London

Berger AN, Udell GF (1998) The economics of small business finance: The roles of private equity and debt markets in the financial growth cycle. J Bank Finance 22(6–8):613–673

Bolzani D, Munari F, Rasmussen E, Toschi L (2017) Technology Transfer Offices as providers of science and technology entrepreneurship education. Paper presented at the international research conference on science and technology entrepreneurship education, Toulose, France, 27–28 April 2017

Croce A, Grilli L, Murtinu S (2014) Venture capital enters academia: an analysis of university-managed funds. J Technol Transfer 39(5):688–715

Darcy J, Kraemer-Eis K, Guellec D, Debande O (2009) Financing technology transfer. European Investment Fund, Adenauer, Luxembourg

Giuri P, Munari F, Pasquini M (2013) What determines university patent commercialization? Empirical evidence on the role of IPR ownership. Ind Innov 20(5):488–502

Lerner J (2005) The university and the start-up: Lessons from the past two decades. J Technol Transfer 30(1–2):49–56

Lerner J (2009) Boulevard of broken dreams: Why public efforts to boost entrepreneurship and venture capital have failed—and what to do about it. Princeton University Press, Princeton

Manigart S, De Waele K, Wright M et al (2000) Venture capitalists, investment appraisal and accounting information: a comparative study of the USA, UK, France, Belgium and Holland. Eur Financ Manag 6(3):389–403

Manigart S, De Waele K, Wright M et al (2002) Determinants of required return in venture capital investments: a five-country study. J Bus Ventur 17(4):291–312

Mason C, Harrison R (2001) 'Investment readiness': a critique of government proposals to increase the demand for venture capital. Reg Stud 35(7):663–668

Munari F, Toschi L (2011) Do venture capitalists have a bias against investment in academic spin-offs? Evidence from the micro- and nanotechnology sector in the UK. Ind Corp Change 20(2):397–432

Munari F, Pasquini M, Toschi L (2015) From the lab to the stock market? The characteristics and impact of university-oriented seed funds in Europe. J Technol Transfer 40(6):948–975

Munari F, Rasmussen E, Toschi L et al (2016) Determinants of the university technology transfer policy-mix: a cross-national analysis of gap-funding instruments. J Technol Transfer 41 (6):1377–1405

Munari F, Sobrero M, Toschi L (2017a) Financing technology transfer: assessment of university-oriented proof-of-concept programmes. Technol Anal Strateg Manag 29(2):233–346

Munari F, Sobrero M, Toschi L (2017b) The university as a venture capitalist? Gap funding instruments for technology transfer. Technol Forecast Soc Change 127:70–84. https://doi.org/10.1016/j.techfore.2017.07.024

Schibler E, McAdam R (1998) Study of the financial intermediary market in Atlantic Canada: KPMG consulting for Atlantic Canada Opportunities Agency

Wright M, Hoskisson R, Busenitz L et al (2001) Finance and management buyouts: agency versus entrepreneurship perspectives. Venture Capital 3(3):239–261

Wright M, Lockett A, Clarysse B et al (2006) University spin-out companies and venture capital. Res Policy 35(4):481–501

Part III
New Methodologies for Capacity Building

The Perspective of an Expert on the New Capacity Building Methodology, Mentoring and Coaching for Technology Transfer

Thomas Flanagan and Elke Piessens

Abstract

This chapter offers a twofold perspective on capacity building in technology transfer. On the one side, it recounts the experience of a mentor in the ProgressTT coaching program and the perception of it by the head of the technology transfer office that received support. It also helps to understand how the cases studies and the new methodology has been used with respect to the several critical areas of focus. On the other side, a number of best practices is reported, that allows to reader to understand how cases can be selected, structured and made instrumental to coaching and mentoring activities within a capacity building program.

An Introductory Note from the Editors of this Book

A useful way to present samples of the best cases and to explain how they could possibly be used for capacity building actions was to ask one senior technology transfer professional and ProgressTT mentor to provide his own personal view and recount his experience in applying the new methodology and ask his mentee to comment on the experience. The chapter has been written by Tom Flanagan, Director of Enterprise and Commercialisation, University College Dublin, Ireland, formerly Director of Hothouse, Dublin Institute of Technology (DIT) (mentor) and ir. Elke Piessens, Director Tech Transfer Office, University of Hasselt, Hasselt, Belgium (mentee). Their contributions provide a fresh perspective on the methodology but are also strongly inspirational as they reveal how capacity

T. Flanagan (✉)
NovaUCD, University College Dublin, Dublin, Ireland
e-mail: tom.flanagan@ucd.ie

E. Piessens
Technology Transfer Office, Hasselt University, Hasselt, Belgium
e-mail: elke.piessens@uhasselt.be

© Springer International Publishing AG, part of Springer Nature 2019
M. Granieri and A. Basso (eds.), *Capacity Building in Technology Transfer*,
SxI – Springer for Innovation / SxI – Springer per l'Innovazione 14,
https://doi.org/10.1007/978-3-319-91461-9_8

building is not necessarily a one-way exercise in transferring or creating technology transfer capacity, but also a two-way enriching experience where also the mentor has the chance to challenge his own views and refine his knowledge about the meaning and the practice of technology transfer.

The selection of cases, clustered around the Critical Areas of Focus (CAFs), is also revealing: the technology transfer environment in Europe is rich with examples of practices that could become a terrific reservoir of cases for any TTO motivated to change and for any policy maker willing to introduce measures to support the technology transfer process and improve the ability of research organizations to create social and economic impact.

1   Introduction

Since I was not involved in the first two years of ProgressTT when the team developed the methodology and the case studies, I can share with you the perspective of someone using the ProgressTT methodology as a mentor. My mentee's comments will appear in italics throughout this section. Along with 14 others from 9 countries across the EU, I was delighted to accept the offer to be involved in ProgressTT and take on the roles of mentor and coach to the TTO at the University in Hasselt, as it showed significant potential for growth.

I was excited to be working with a strong team and a very capable leader Elke, who had a great interest in learning best practices from across Europe and was ambitious to improve the performance of the team and grow the operation in line with the expectations of her senior management.

> And I can add as the mentee that we felt that teaming up with the right Key Expert would support us greatly in analysing our situation and setting out an action plan to help us to get to a higher level. At the moment we decided to participate, we indeed needed an experienced outside view to see our current pitfalls and the opportunities, and support us to translate that into a 5 Year Strategic Plan that could be operationalised for future growth.

From a mentor/coach perspective here is one of the first ingredients of the secret sauce that makes the ProgressTT methodology so good: a lot of work goes into assessing the candidate TTO, understanding their capabilities and what was holding them back from greater performance. The second ingredient is the matching of mentor with client. In my case the operation I run is less than half the size of the client's and the research funding is less than half and they do substantially more collaborative and contract research agreements with industry but we had twice as many spin-out companies and ten times as many licences per year.

So why were we such a great match? I think the closeness in size of operation and yet the significant differences in performance meant that when we compare in detail how we each operate we could easily see the different practices and opportunities to improve performance.

Also very important was the fact that we spoke the same TT language and that the coach knew exactly the framework for phases of growth for TTOs. Next to that, I think the fact that we connected as individuals, open for mutual interaction, and could communicate easily also helped.

It is interesting as well that when we brought our teams together; they also had a lot to share with one another and gained a lot from the engagements. In our case the mentoring/coaching engagement was just the start of what I expect will be an ongoing relationship between our teams that will survive and thrive well beyond the ProgressTT project timeline. And that sustainability where the engagement and the value creation continue beyond the lifetime of the EU project is a real achievement for the ProgressTT team.

Our two teams were eager to share experiences and learn from each other. I agree completely that our relationship thus will go on.

2 Critical Areas of Focus

The Mentor/Coach handbook and the webinar training were very useful to ensure that all mentor/coaches understood the methodology and were using the same language and templates. It was well done and left few questions unanswered while still providing a lot of freedom for the mentor/coach to interpret the guidelines to specifically meet the needs of the clients.

The training took us through the assessment that was being used to select the candidates and it was very well structured. The selection of four Critical Areas of Focus was novel and very productive. There are more than 10 processes involved in technology transfer (proposal support, invention identification, patenting, managing Proof Of Concept (POC) funds, developing marketing materials, promoting the inventions, identifying target licensees, negotiating, managing contracts, finding CEO's for spinouts, spin-out formation, collecting and distributing royalties, etc.) so to consider just four Critical Areas and focus on two of them was very productive.

Left to our own devices most TTO Directors and mentors would have spent the first engagement trying to decide on the areas of focus and depending on the individuals involved, it may have been very challenging to not attempt to tackle every area. To focus on just two of the four major critical areas meant that the energies of the TTO and the mentor were not spread so thinly that real impact could not be achieved.

If the first and second ingredients were the assessment of mentees and the matching of coaches, then the third ingredient of the secret sauce that makes the ProgressTT methodology so good came from the choice of Critical Areas of Focus. We focused on areas to improve where we would move the TTO from "good" to "great", and did not focus on areas where they had no capability at all. Areas where there is no capability are usually better addressed through recruitment of new skills rather than training, mentoring or coaching. This was a clever although non intuitive approach. Most people's first instinct would be to focus on the areas of greatest need rather than the ones that could benefit most from mentoring/coaching.

> Also for me the approach of using the Critical Areas of Focus was new. Because it was limited to four areas, it was simple to capture, but yet covered all relevant TT processes. We selected two areas where indeed we had already capability and wanted to grow. But during the process, automatically also links and improvements on the other non-selected areas were discussed, not in-depth but enough to tackle those ourselves within the Strategy Plan and actions.

3 Case Studies

It often seems that we all do the same things in technology transfer; we engage academics; find, protect and market interesting intellectual property (IP); if successful we find a licensee, negotiate a licence and collect royalties; or we help the academic or external business partner launch a start-up companies based on the IP. However in reality, we don't all do the same things and we certainly don't do things the same way across Europe or around the world. The differences are often subtle but have great impact.

When I started out in Tech Transfer, setting up the first ever TTO office in Dublin Institute of Technology (DIT) I was avid to learn all of the best practices (Flanagan 2017). I remember going to see the Oxford and Cambridge TTOs and incubators and I remember thinking as they talked about holding networking and showcase events for example that we were doing these in DIT too and that there was very little new here. It was not until I dug deeper that I realised that while the menu of activities was largely the same, the frequency with which they had events was an order of magnitude bigger from what we had underway in DIT at the time.

Like restaurants, while the menu may sound the same, the experience can be quite different. So while it may sound like we are all doing the same things, we are in fact not, we sometimes do significantly different things to achieve the same results or when we do the same things we can often do them in very different ways and this results in very different performance and impact. The case studies in ProgressTT illustrate this further as they tackle the Critical Areas of Focus in very different ways.

> The experience of the coach was the first important driver at the start of our journey: from his own successes but also mistakes, he analysed quite fast our current situation based on the CAF approach: which TT processes do we have in place and how do we do them. With some right placed questions, he each time addressed quite straightforward our strengths and weaknesses, got us to think about them and find possible solutions and actions. In this phase of the coaching, the case studies indeed provide extra viewpoints and solutions.

The common problem/solution structure of the case studies together with the ProgressTT classification aligning them to the Critical Areas of Focus is a great help to enable the mentor and client find alternate solutions to common problems. The commentary at the end of each case highlights the key learnings that can be used elsewhere. But can these learnings be easily replicated? I believe so, but it requires a lot more detail than can be presented in two page summaries. It requires more detailed understanding of the cases. It may also require mentoring to ensure that the cases are fully understood and adapted to the local needs and coaching to ensure that the initiatives are implemented and seen through to fruition.

The common factors in the cases selected are that someone in the TTO identified the problems they were having and took actions often several initiatives to resolve the problems. Like every story of success it belies all of the stories where someone identified the problems undertook initiatives but still failed either through incorrect diagnosis, incorrect initiatives or poorly executed ones. Often the environment is very different and can sometimes change positively or negatively as the initiative is implemented. So the real opportunity here is using the case studies as sign posts to practitioners who can share the additional detail needed to compare your environment with theirs and give you an understanding of the critical contributing factors that led to their success.

> During the coaching, we selected several cases which we found applicable for our situation and that tackled an issue we were also struggling with. As mentioned above, the 2 pages were not enough to apply the problem/solution to our own situation, so the possibility of getting in touch with the owner of the case by phone or in a meeting, were of high added value.

> To make good use of the cases as a mentee, it is important to know if the TTO in the case is comparable to yours in tasks and services, or not. In this respect, a possible improvement on the template used for the cases, could be that the legal structure of the TTO is more clearly mentioned (incorporated within university or independent organisation), which processes they facilitate (research contracts, patents, licenses, spin-outs) and how they are funded or being evaluated.

4 Use of the Case Studies

We will discuss the challenges in each case presented in this book from the perspective of a mentor/coach and client who are looking to adapt them and replicate the impact in their own environment.

4.1 CAF 1: "Scouting Ideas/Technologies from the PRO and Incentivising Researchers to Disclose IP"

There were several excellent case studies to draw from to address the CAF1 issues of finding more inventions and incentivising researchers to disclose IP. Specifically we looked at the University of Ghent and the University of Copenhagen Spin-out cases included here which both referred to establishing scouts. The TTO at the University of Hasselt already had business developers in the Research Centres, however they had not been specifically asked to scout for IP or to meet invention disclosure targets and were more focussed on developing collaborative research. Once they were asked they were more than happy to do it and saw lots of opportunities.

The University of Copenhagen start-up case not only described the introduction of scouts but also had a number of other incentives to make scouting more effective, including the use of scouts to look for cross-over research opportunities and engagement with senior management giving senior management a more inclusive view of the outputs of research across the university. As well as giving the scouts greater visibility in the organisation, this also served senior management well in helping them understand where the strengths were in research and innovation across the university.

> The Copenhagen case also shared a number of schemes to promote entrepreneurship and inventorship through competitions, awards, etc. which we will also implement.

4.2   CAF 2: "Assessing IP Potential, Validating Technologies and Incentivising for Commercialisation"

The patent portfolio management process used by my client TTO was to only file patents where they expected they would receive a return covering the cost of the patent filing within a year for initial filings or a further 18 months for a PCT. It was quite restrictive to be looking for such an immediate return on investment. Consequently the number of patents filed was low for the level of research they had. You might expect to see 2–4 patents filed each year per €10 M of research (Directorate-General for Research and Innovation 2013). Having the cases and access to experts across Europe meant we could easily survey this ourselves when we engaged with the case owners. In fact we learned when we engaged with the University of Liege and the University of Leuven that their processes were very similar to our own in DIT Hothouse. We each patent as many inventions as we can within budget and try to licence them within the first year. We make a decision on PCT for a further 18 months patent coverage if we think that we can get a licence within this period if not we generally let the patent die or offer it to the researcher. This alternative process was taken into account by my client TTO and they are making a proposal for increased patent budget to better serve their commercialisation opportunities.

The process of marketing and finding licences at my client TTO was also limited. They didn't create non confidential summaries of their inventions but just referred generally to their research expertise. The process of marketing technologies was addressed in the "Ideas exchange platform at Cracow University of Economics TTO" case where they have created a common web platform to highlight their offerings to industry. We have done something in Ireland where the offerings from the Universities and Institutes of Technology are presented on a common website www.knowledgetransferireland.ie run by Knowledge Transfer Ireland, the state agency responsible for enhancing industry academia engagements. But good TTOs do a lot more than put their technology summaries on a website and wait for a call. Most of us actively target and engage potential licensees to see if the technologies are of interest to them and if not what are they interested in which might create new opportunities for collaborative/contract research or consultancy.

One of the other big issues identified by my client TTO was with POC funding. The issue was that while they had funding for several projects each year, they got more applications from researchers than they could fund so some researchers whose projects were not funded were dissatisfied by the communication around the POC and the decisions made by the TTO on funding. In reality some of the proposals were very early stage and would not be progressed to commercial interest with the small sums offered in the POC. So how could we use POC funding to create a more positive engagement with researchers?

There were several good case studies around POC funding. We decided to look at the University of Liege case included here and the UCL case that was available on the ProgressTT website. In the case of the University of Liege they have developed a very thorough and sound series of questions along 6 categories (technology readiness, market conditions, patentability, patent status, human resources and finance) and when the questions are answered from 0 to 6 you can draw a spider diagram showing the current state and another spider showing what you expect at the end of the POC. In most cases the University of Liege are looking for the POC to advance the technology from Technology Readiness Level TRL 4 to 5. In the case of life science projects they would fund a POC up to TRL level 3 to 4.

It was also very instructive and comforting to understand that the University of Liege TTO appreciates that the answers to the questions are subjective and not every two licensing executives would have given the same project a five for example for patentability when they started using it but that overtime and through the use of peer reviews their assessments and ratings had become closer. It was also worthy of note that the assessment was being used as a relative measure not an absolute measure to select one project over another and to identify the improvement if any against the criteria that should result from completing the POC project.

The TTO at the University of Liege also manages the budget of the POC so that it is tightly controlled in relation to the milestones of the project as they progress toward commercialisation and they terminate projects part way if the progress expected is not achieved. This is different from other cases where the state agencies give POC funding to the researcher up front and although there are interim milestone reports the researcher is generally free to manage the funds with less supervision.

In the case of UCL, the business development people supporting the research centres develop the proposal with the researchers and if it is funded they take the lead and manage the project with the researcher supporting them. They typically make the proposal to the TTO Director if the amount is small (less than £25 k). The proposal will have had peer reviews by business developers and TTO staff to see if the project will bring the technology to a state where there could be commercial interest in a licence or it could attract investment for a spinout. This approach makes the decision more commercial and it is easier to justify to the researcher if their project is not funded. My client TTO now plans to adopt this approach.

The second case CAF2, included here, from University College Dublin (UCD) was also of great interest to my client as they were looking to increase the number of successful spin-outs they could achieve. I knew the UCD programmes that were included in the case study here and we run similar programmes at DIT Hothouse. The case study describes at a high level the topics covered and it outlines the structure of the programmes. They are an excellent starting point. The case study from the University of Linköping from the ProgressTT website was also of interest. We had a call with the University of Linköping and learned that it is run from the Business Faculty and has few requirements for getting onto the programme, just an interest in the programme is sufficient.

The UCD programmes and our own programmes at DIT Hothouse would both have very tight entry requirements. We choose to be very selective on the front end selecting 1 in 5 to 1 in 10 applicants and I know of some programmes in Canada, where the ratio is reported to be as high as 1 in 40. We give feedback to those that don't make it so that they can do some further work to prepare better before being selected in the future. The tighter selection means that we have stronger candidates on the programme and there is better peer learning. We can also give each CEO more attention and more tailored feedback focussed on each of their individual business challenges.

Our programmes are focussed more on "enterprise development support" i.e. helping the entrepreneur launch a successful business than on "entrepreneurship training" i.e. providing generalised training on launching a business. It is a subtle but very significant difference. We don't measure success by the numbers interested in the programme or that have gone through the programme. We measure success by the number of businesses that are still alive after 5 years and the number of jobs they have created and the investments raised. The development of such a programme for my client is now in their strategic plan.

> CAF 2 was in our Individual Action Plan and as you can see above we have learned a lot through the case studies and from our mentor and we now have plans to make some significant changes in this area to enhance our performance.

4.3 CAF 3: "Accessing Finance and Interacting with Financial Stakeholders"

One of the big challenges most TTOs face is how to launch Life Science start-ups that require substantial financing and are very high risk and my client was no different. Many of us rely on finding a good CEO that can build the team and sell the business proposition to investors but they are hard to find, as are the investors willing to fund such early stage projects. The approach taken in the University of Cardiff case study is inspirational as it improves the probability of success by having external professional investors manage a portfolio of such start-ups through the early stages.

The University of Cardiff developed a strategic relationship with an external investment company, Fusion IP, now part of the IP Group plc, whereby Fusion IP ring-fenced £8.2 m for investment in a portfolio of University of Cardiff spin-outs called Biofusion and gave the university 30% equity in the portfolio. Having professional investors on board at the early stage, seeding the initial investment, building and mentoring the management teams and leveraging their investment to attract other investors proved to be very successful for all involved. It was certainly successful for the 20 start-up companies involved which attracted £50 m in investment. One of these companies subsequently floated on AIM (Alternative Investment Market, subgroup of London Stock Exchange). It was also successful for Fusion and for the University which was able to sell shares in the IP Group plc and see a return.

The case itself is inspirational but the real value comes for a mentor and mentees to have access to the key players involved in this case to ask them the hard questions about the selection criteria for candidate start-ups, the selection of CEOs and start-up teams, the gate reviews, the challenges in getting other investors on board, the actual return to the University on the portfolio versus going it alone with individual projects, the preparatory work required of the TTO, etc. This is very powerful and enables other TTOs to see if and how they might emulate or improve on the Cardiff model.

The other case provided by the University of Bergen, included in this section, shows that the portfolio model can work for a range of technologies where large investments are needed, not just Life Science projects. It also shows that while the portfolio fund was initially focused on a region and had no government investment, it was attractive enough to secure investors from all around Norway and abroad.

This case provides a greater understanding of the role of the TTO in developing the projects to the stage where the portfolio managers would select them for investment. The TTO team initially act as interim CEOs developing the projects to a stage where substantial investment is needed, typically 20% of these projects are selected by the portfolio management company NIK (Norsk Innovasjonskapital AS) and shares are swapped so that the University has part ownership in the portfolio company rather than the individual start-ups. The University then gets a return when the portfolio is sold on or as it manages an exit for any of the university spin-outs.

Again having access to the players to understand the stage of development required before NIK acquires the company, the pipeline of potential projects available each year, the competition NIK has versus other private or VC investors, the negotiation of share value in the transition, the impact that the acquisition has on retaining or changing out the existing management team, the ROI for a university versus going it alone with other investors, are key to universities and VC's deciding to implement a similar scheme.

Both case studies share approaches to attracting large scale investments into high risk ventures by professionally managing them as a portfolio. The differences are in the stages at which the acquisition of the start-up takes place, earlier in the case of the University of Cardiff. What is very significant about the University of Bergen approach is that it has raised more funds than the government backed seed investment funds. This says a lot about the value of professional management and pooling of opportunities to de-risk the opportunities for institutional and private investors.

Although we were not focussed on this in our intervention, both cases are inspirational and we may well implement a similar initiative in the future.

4.4   CAF 4: "Securing TTO Staff Skills and Organising the TTO for Optimum Growth"

TTO staff require a great many skills; they need to understand the technologies to be able to differentiate them from what has been done before so that they can identify unique selling points and decide patenting strategies; sometimes they draft patents or advise on drafting strategies; they need to be able to conduct market research to identify all of the possible applications, markets and target companies for their technologies and then identify individuals within these companies and approach them; they need to be good at selling or persuading potential licensees or CEO's to engage in contract or collaborative research, to take a licence or form a company around a technology; this often involves preparing a business case or business plan or at the very least a lean canvas for the opportunity, which in turn involves understanding the competitive and regulatory landscapes; they also need to be able to negotiate commercial arrangements; sometimes draft the agreement or engage and manage legal experts to draft and redraft agreements until the deal is done; and finally they need to follow-up with licensees and researchers to ensure that both academic and industry clients are satisfied; that royalties are paid and distributed fairly etc. and if they are smart they also look for repeat business.

The skill set is broad and needs to be deep as well in order to win over the trust of both the academic and industry clients. It requires technical skills: science and/or engineering; patenting skills; market research; technical marketing; project management; selling; negotiation; and legal skills. It also requires a range of soft skills: building rapport; active listening; mediating disputes; communicating effectively; presenting; running meetings; persuading; building long term relationships at every level in their own organisation and in diverse industry client organisations: multinationals; Small Medium Enterprises (SMEs), with entrepreneurs, angel investors and Venture Capital (VC) firms and across government agencies and sponsoring organisations.

The skill level required is very high so it is not surprising that most TTOs are very interested in understanding how best to recruit and retain this talent. The two cases provided in this section share some best practices. The case of the Institute of

Organic Chemistry (IOCB) TTO illustrates the launch and transition of a TTO from a staff of two providing commercialisation advice to the research community which incidentally didn't result in any real commercialisation to where it evolves to hiring additional project development staff to progress the commercialisation projects to fruition.

IOCB TTO started with a commercial CEO and a patent expert but once they secured additional funding from the Czech Republic Technology Agency for the Centre for the Development of Original Drugs, a consortium of 6 businesses and 3 universities, they were able to hire the project development team. Each project manager manages 3–5 projects per year. They also have seed funding of 37–147 k€ per project. Staff training is a mixture of on the job and external training. They continue to be supported financially by their own organisation IOCB and by the consortium but hope to be self-financing based on licensing income in the future.

The Unitectra case also included here is that of a more developed TTO which provides TTO services to three Universities. It describes the qualifications (up to and including Ph.D. level) and the commercial experience (5/6 years in business development) that they look for in recruiting and it highlights the need for on the job training with mentorship over a two-year period before the person is really effective.

The Unitectra case also describes the structure of their organisation which supports three universities in the region of Basel, Bern and Zurich. It is interesting to understand how they have avoided the pitfalls that normally are associated with a common pool of TTO staff serving multiple organisations. They started by sharing costs on the basis of size of organisation supported but evolved to a pay per use model. They also resolved the issue of connecting with researchers in different locations by spreading their staff across the different campuses to increase proximity and familiarity with the research teams. Furthermore part of the TTO team rotates between the campuses to ensure quality and commonality of approach.

IOCB TTO addressed the concerns about potential bias for one university over another by being as transparent as possible and having a board that has two representatives from each university. They are also very proud of the retention of staff with half of the team having stayed for more than ten years. They attribute this to flexibility in the role and offering competitive salaries with industry.

From a mentoring perspective there are lessons to be learned from both cases. The first case involves the establishment of a TTO and the need to have a critical mass of projects and project development staff to be successful. It also highlights some approaches for securing the funds needed to establish and grow a TTO. The second case shows the benefits of aggregating still further and providing TTO services across three universities for economy of scale. In both cases very experienced staffs are required and their skills are continuously developed so that they are current with best practices.

My client TTO was interested in setting up their organisation and their processes for significant growth. The two cases discussed above provided good insights into the qualifications, experience and the training needed to launch and grow a TTO. The first case involved the establishment of a TTO and the second case shows the benefits of aggregating still further and providing TTO services across three universities for economy of scale. However since my client TTO was already well established and had sufficient critical mass that it could remain stand alone, neither of these cases provided all of the answers to the challenges of growth facing them. So we also looked at the case of the University of Leuven available on the ProgressTT website.

The Leuven case write up gives an overview of all of their operations but we were interested in learning exactly how they were structured and how they had grown. A one hour discussion revealed so much more than could be condensed in the two page case study. The office had grown from 12 in 2000 to 90 in 2017. The structure included an IP team of 8 licensing executives and 2 legal experts. Each licensing executive finds inventions, writes patents, finds licensees and negotiates licences. Others in the team are involved in HR, Finance, contracts, consultancy, EU funding, spinouts and student enterprise support is run by students for students.

The licensing executives have very important roles in terms of completing their own market intelligence before engaging external experts, venture capitalists and potential licensees. Similarly from the University of Liege engagement we learned that their licensing executives and business developers have on the job training from more experienced licensing executives on how to prepare for going to market with their technologies.

> The four cases discussed here together with the Swim Lane Organisational Analysis that is discussed later provided the necessary information on structure, roles and responsibilities to help us redesign our organisation for growth.

5   Mentoring and Coaching

Mentoring involves sharing your expertise with the mentee while coaching involves giving feedback on the behaviours the coach observes. In ProgressTT we used both mentoring and coaching to very good effect. For.the client the intervention was very much a journey of discovery and commitment.

There were issues to discover throughout the intervention. It was important to ask lots of questions and not to make assumptions. There was also a need to ensure commitments were met throughout the intervention. In my experience a good way to achieve this is to first model the behaviour you seek in others. In this case make commitments and keep them no matter how small. Also look for commitments from the client and follow up on each one at the early stages of the engagement so that the client knows that you have a high standard when it comes to commitments.

> Our coach was very good in asking the right questions to make us see our strengths and weaknesses. He didn't point them out himself, but got us to phrase them ourselves, which was stronger because we made the analysis and conclusion ourselves, which was a strong empowerment towards ownership.

The coach was indeed on top of the commitments made via e-mail, conference calls and visits. This was a strong motivator to work on it and deliver quality in time. What worked very well, was that the coach often gave us a good starting point, and we could work further on it to score.

Some of the other subtleties of the ProgressTT approach that worked well were the time limit of six months for the intervention. It is long enough to make a difference and short enough to let the client get on with their business and not become dependent on the mentor/coach. If the mentoring/coaching had been spread out over a year the momentum to make change would have been lost as the daily demands would likely distract the client and reduce the focus on making immediate changes.

There are however good arguments for doing a follow up health check a year later to see how much progress has been made and to tweak any behaviours or processes to improve performance. This would also give the client a future goal that would help them keep up the momentum of the initial intervention.

I fully agree. No doubt further coaching (e.g. a check-up each 6 months) would stimulate the client TTO to take further actions on making changes and stick to the Strategic Plan, and not get caught up in daily business.

5.1 The First Meeting

In preparing for the first meeting in Milan with the assigned client TTO Director, we both checked each other's linked-in profiles and used the contacts we each knew to understand who it was we were meeting, their backgrounds etc. to guess what their expectations might be. I also reviewed the TTO's self-assessment diagnostic highlighting weaknesses and areas for improvement and looked at the Individual Action Plan (IAP) template and the milestone template that we were going to fill together at our first meeting.

ProgressTT had arranged a social get together the night before the official first meeting which allowed us to get to know one another socially. It was very helpful to get to know one another in this way before getting down to business the next day.

On the day, the ProgressTT team provided an overview of the process and highlighted the broad range of cases we could reference. We then met the client to understand the real needs well beyond the diagnostic. The pie chart in the IAP showing the relative importance today and in five years of supporting academics, engaging industry or focussing on generating revenue from TTO operations was a great conversation opener. My client chose the biggest focus to be on industry engagement followed closely by supporting academics and had only 10–15% focus on generating revenue. Fortunately for my client they later found out that their Chancellor shared the same view. This focus is common for most TTOs I know in Europe.

Understanding the client's timeline for change and the major changes they sought to achieve in the six-month's engagement was critical. The use of active listening—repeating back what you heard, asking open ended questions, making no assumptions, but questioning everything in a friendly way was important. It was also helpful to compare their operations process by process against my own operation to see where they were different. This helped us decide on the Critical Areas of Focus and helped the client feel comfortable that the mentor selected might be able to help them.

It was interesting that before this meeting my client had considered enhancing the TTO skills in isolation not taking into account the larger business development team they had located in the research centres. The business developers in this case had been treated separately as outside of the TTO. It was important to reframe the discussion to include them. We agreed that for the purpose of the ProgressTT project we would treat the seven TTO staff and the eight business developers as one team.

Reflecting on this change, treating what was two discrete teams as one, meant that we were going to have to create a shared vision of targets for the team and address any gaps or overlaps in their operation as a team. The number of inventions disclosed and the number of licences completed each year were low for the level of research (One would expect 1–2 licences per year per 10 m€ of research, (Directorate General of Research and Innovation 2013), so we agreed that we should use the ProgressTT project to develop a five year strategic plan for the TTO and business developers. Developing a plan together would create a shared vision, shared targets and a stronger commitment to the changes we sought.

We would begin with a stakeholder analysis (milestone 1) to understand stakeholder expectations, then do a process and organisational analysis (milestone 2) and bring both together in the five year strategic plan (milestone 3). We would include presenting and getting approval on the plan as the last milestone (milestone 4). These milestones would address the major strategic issues but there were also operational issues that needed addressing. Comparing how they go about finding inventions, how they manage their patent portfolio and how they market their technologies to potential licensees were all very different from the way we operate in DIT Hothouse. Together we selected several case studies to address their two Critical Areas of Focus: CAF 2 "Assessing IP potential, validating technologies and incentivising for commercialisation" and CAF 4 "Securing TTO staff skills and organising the TTO for optimum growth".

We completed the first meeting by planning out the engagements we would have; the three onsite visits and the one hour conference calls that we would have every week. We packed most of the work including the onsite visits into the first three months so that the last three months could be spent refining the five year strategic plan and getting approval on it. It was best to schedule it that way to build on the initial enthusiasm that usually accompanies the start of a project and to allow for unforeseen circumstances that might delay elements of the project but still allow us to finish the project successfully within the six months.

The above summary reflects precisely our first meeting. It was exciting to finally get to know our coach in person, and to see if we matched. I can only add that we certainly could use an extra half day to have more time to start even better off with our coaching track. Maybe an introductory webinar on the beforehand with the general information on the project, actions, expectations and timeline, would even enhance the efficiency of this first 'face to face' meeting itself.

5.2 The First Onsite Meeting

The team planned fourteen internal and external stakeholder interviews with: the Chancellor, Vice Chancellor, directors, researchers, business developers; external partners and clients including venture capitalists, development agencies etc. Scheduling the interviews took longer than anticipated but the learnings would be significant. I also supplied a questionnaire that I had developed in DIT Hothouse for assessing researcher client satisfaction with the TTO that we would send to a broader range of researchers. A questionnaire like this can achieve three objectives: you can create awareness of the TTO services by asking researchers to check what all services they have used of the full list of TTO services; you can also get honest feedback on what they think of the delivery of these services and get suggestions for improvement; and you can tailor a question that once completed gives you marketing material for your TTO that allows you to promote that for example 90% of researchers surveyed thought the TTO services were very good or excellent.

> The combination of interviews on each level of our organisation, internal and external interviews, together with the broader TT survey, gave us a good view on our reputation, services offered/used and the expectations of our 'customers'. This was an eye-opener and a good starting point to make our new strategy plan realistic and in line with the expectations of our target customer group.

In the three-day visit, the TTO Director and I conducted interviews with the Chancellor, Vice Chancellor, Head of Research, the head of one of the most progressive research centres, and a business developer. We also met with each of the TTO team individually and with them as a group.

The TTO Director thought that it was very useful to have an external mentor/coach to ask the harder questions of the Chancellor and Vice Chancellor about the reputation of the TTO in the business and research communities and what needed to be improved.

> The top management of an organisation is logically always striving to increase the output. With respect to TT output, we can on the one hand increase efficiency in our processes, but at a certain point, we have to increase capacity if we want to raise output further. If that investment in human capital cannot be made, the management has to agree to make choices and focus on certain targets. It was good to have an external expert confirming this.

It was also beneficial to the TTO Director and to individuals in the team to have confidential one on one mentoring discussions on operations, processes, organisational design, performance, management and leadership. While these discussions were treated as confidential, it was good to use the insights to develop and suggest enhancements to benefit the team and the individuals.

The confidential one on one discussions were indeed very useful and we are now scheduling them regularly as part of our normal operations.

The stakeholder meetings told us what was working well and what needed improvement. There was positive feedback on the value of the business developers in each of the research centres, an initiative they had undertaken two years ago. The business developers were really adding value in developing research collaborations and driving up research funding and industry engagements. The constructive feedback they received in the interviews was that the TTO was less visible in its impact in the wider community and there were perceptions that they were responsible for nearly always saying "no" on POC funds as we discussed in Sect. 4.2.

The stakeholder meetings were also an opportunity to promote the TTO function internally and externally by telling the stakeholders that the KTO (TTO + Business Development) had been recognised by ProgressTT as one of only a few TTOs across Europe that had showed great potential for growth and were therefore selected and being supported by ProgressTT.

During the coaching project, the KTO became a real integrated team, ready to grow further together.

While we were together onsite we also had a call with the TTO Director at the University of Linköping (one of the case studies) to understand how they were encouraging researchers to start businesses.

5.3 The First Milestone

The TTO team had engaged all of their key stakeholders, listened to them and they were planning to take action to enhance performance. They were now able to complete the milestone 1 report on stakeholder analysis.

They summarised the key actions as follows:

- They needed to develop a few key targets and deliver on them rather than only review a great many KPI's at year end.
- They needed to increase the number of invention disclosures, then develop non confidential summaries and market them better to industry.
- They also needed to develop greater engagement with the ecosystem of incubators and businesses in the region to find good CEO's and get support for start-ups.

Through the stakeholder analysis, it was clear that we have the team, processes and KPIs largely in place, now is the time for focus and actions: just do it!

5.4 The Second Milestone

To achieve the second milestone of process and organisational analysis, I visited the TTO again and we held a workshop onsite with the full team of TTO and Business Development people to talk through their processes and operations using the Swim Lane tool.

The Swim Lane tool is a matrix. Across the top, the various processes are listed from initial proposal generation, to invention disclosure, to patenting, to marketing, to licensing, to royalty distribution etc. And along the vertical, the various roles are listed: TTO Director, TTO staff, Legal, Business Developers, Marketing and Administration. Each square of the matrix is then filled in by the team deciding who has the lead on it and who else is involved.

The use of the Swim Lane tool is very effective in seeing the gaps and overlaps in operations and helps everyone understand who the lead on each operation is. Once this is agreed the next step is to set targets for each lead which in turn leads to the five year strategic plan targets.

> The Swim Lane exercise was extremely useful for the entire TT team, since we have 'business' tasks as well in the central TTO as with the local business developers. The persons involved in the same process needed a clearer view on responsibilities, who is in the lead, and moments of handover. This transparency will prevent processes stopping when nobody 'knows' who is in charge of the next step, and will avoid persons involved in the same process being frustrated when they have different expectations towards each other.

5.5 The Third Milestone

In a third three day visit, I reviewed a proposal by their start-up person on how they would tackle start-up creation and provided guidance on setting priorities around finding great projects and great entrepreneurs to lead them. I also delivered an Invention Identification and Market Intelligence Workshop to the TTO and business development teams based on what we do at DIT Hothouse. We also had a call with UCL to review their approach to POC funding and connected with the University of Liege to review their Innovation Readiness Levels tools for assessing patentability and research commercialisation progress.

To achieve the third milestone we developed suitable year over year targets for five years for Inventions, Patents, Licences, Spinouts and Research Agreements and reviewed and finalised the milestone 1 and 2 reports on Stakeholder Analysis and Organisational Design. We also mapped out the key components of the five year strategic plan.

> The very target-focussed, commercial and organisation-wide thinking of our coach was an eye-opener for me, especially the exercise to start from the global research budget of the organisation towards expected TT output, and go for a limited set of targets (increasing), chosen out of the large number of KPIs (for monitoring). Next to that, the use of committed targets (towards management) and stretch targets (as an internal motivational tool for the team) was a handy tip. Our ambitious coach also helped us in setting targets by showing us

how to set targets based on the average performance of TTO's across Europe, setting the number of licences and spinouts per €10 M of research etc., but then he inspired us to do more with his comment. "Of course Elke, we don't do average".

Seeing the TTO take the lead on setting year over year increasing targets for themselves on licences, spinouts and research agreements had the follow on effect of raising the question for the Chancellor, Vice Chancellor and Heads of Research Institutes of what they were targeting and committing to for research growth year on year. It was a new approach that inspired senior management to begin to set ambitious but achievable targets for themselves in research and innovation.

The exercise also showed higher management that all targets throughout the organisation are linked to each other and often are dependent (research targets and TT targets and BD targets). Higher management need to coordinate the setting of shared and realistic targets for the entire organisation, breaking them down to individual coherent targets at all levels.

The TTO Director developed a presentation summarising the feedback received in the stakeholder interviews as well as the learnings from the organisational design workshops and the use of the Swim Lane tool. They also set out following consultation with their team, the committed and strategic key target metrics for the next 5 years and highlighted a number of concrete actions, many of which were already underway, to achieve these ambitions. The quality of the five year strategic plan was excellent.

More than half of the TTO and business development team visited DIT Hothouse in Dublin to review their 5 Year Strategic Plan with the DIT Hothouse TTO and Incubation teams and receive feedback from not just myself, the Director, but also the Licensing Execs, the Marketing Exec, the incubation team and the Administrators on the practicality of implementing their plan. There was also time for breakout sessions where each group Licensing, Marketing etc. paired off and had an open and frank discussion of the challenges we each face and the lessons we had each learned.

The engagement between the two teams was excellent. There was great interest in sharing best practices. Our visitors learned first-hand the practical aspects of committing to and delivering on targets in a strategic plan.

Introducing the two teams to one another created the opportunity to have follow-up one on one conversations to answer the questions that the licensing exec, business developer and marketing executives have, through a buddy arrangement, without the oversight of directors. This will have ongoing benefit for both teams as new challenges are faced in each location.

Both teams were eager to learn from each other since within their own organisation, no other 'academic' person is doing similar work. Moreover, since we are (not yet) doing business together, they could be open on their own strengths, weaknesses, problems and tools, from both sides. During the Dublin visit, we already exchanged a lot of ideas, tools and opinions. A follow-up visit to Hasselt of the Senior Licensing Executive of DIT is already planned, and we expect that mutual visits or conference calls will be planned regularly to continuously learn and challenge both teams.

Having done the work of stakeholder interviews, organisational redesign and setting targets for the 5 Year Strategic Plan and having presented it to robust critique from a team that is knowledgeable and opinionated, the TTO Director was well positioned to present it to their Chancellor and senior management team for approval.

5.6 The Fourth Milestone

The 5 Year Strategic Plan had been socialised with the team, and some of the Research Centre Directors with good responses all around. It was then presented to the Chancellor and the senior leadership team of the university and approved. The Vice Chancellor commented:

> The ProgressTT intervention was very timely as the university is developing a new Policy for Research & Innovation for the next 5 years. It was a great opportunity for our tech transfer team to learn best practices from across Europe, and develop a 5 Year Strategic Plan that will complement and support the ambitions we have in our 5 year Policy Plan for Research & Innovation.

One Research Centre Manager commented:

> The TTO Director has engaged with the Research Centres and developed year on year targets for inventions, patents, licenses, spin-offs and research contracts for each Centre which will feed into our Policy for Research & Innovation and increase our collective commitment to excellence and growth.

Others commented that the TTO Director had done an excellent job engaging all of the key external and internal stakeholders to understand their expectations and get feedback on the current operations. They have taken this feedback to heart and have already implemented a number of strategic and operational changes that will make a big difference to the performance of the group.

The TTO Director was also praised by their staff for bringing together the TTO team and the business development team, engaged them in clarifying their roles and responsibilities so that they now had an organisational structure that is fully functional and is well developed to enable and support significant growth.

> I am very proud of our entire TTO and business development team for the way they stepped up to the challenges and have fully used this opportunity to significantly enhance their performance which I have no doubt will greatly increase the impact our research can have over the next 5 years.

6 Beyond ProgressTT

Time will tell if the commitments made in the 5 Year Strategic Plan are delivered. The team is certainly better aligned around targets, they have the tools to increase invention disclosures, market the inventions better, close more licences and launch more start-ups. They also have greater clarity of roles and responsibilities and their year on year targets are aligned with their senior management's ambitions so it is now a case of *"just do it"*.

Their "moments of truth" will come in the next six months as they try to achieve the higher targets. There is a big difference between knowing what to do and doing it successfully. The biggest worry one might have is that they could fall into the "paralysis by analysis" trap, still looking for additional best practices to emulate rather than focussing on delivering and learning by doing what they already know needs to be done.

It is easy to see the strategic changes as they happen fast; approval of a 5 Year Strategic Plan, organisational changes etc. It takes more time to see the operational and cultural changes that are required to give meaning to the strategic commitments, however I am pleased to say that my client TTO has already changed their invention disclosure form and processes to make it easier for researchers to engage. They have also started to develop marketing materials for each invention to be progressed and are identifying target licensees. They are also seeking increased patent budget to strengthen their patent portfolio and change their patent management. They have developed a new approach to managing POC funding. And they are beginning to market their new business opportunities to find external CEOs for their start-ups. So time will tell but they are off to a really good start and the future looks very bright. They may well become a best practice case themselves!

References

Directorate-General for Research and Innovation (2013) Knowledge transfer study 2010–2012. [online] Available at: https://ec.europa.eu/research/innovation-union/pdf/knowledge_transfer_2010-2012_report.pdf. Accessed 1 July 2017

Flanagan T (2017) The evolution of technology transfer in Ireland 2006–2016: a practitioner's perspective. Small Enterp Res 24:1–10

Selected Case Studies

Marcello Torrisi

Abstract

This chapter provides some of the case studies which were included in the "Best practice library" of the Progress-TT project. This library was aimed at enabling wider dissemination of best practices in the field of technology transfer, through their use in the training events of the Progress-TT project and online public availability of the case-studies in the project website. The case studies included in this library are categorised in four "Critical Areas of Focus" (CAFs) identified as highly relevant to the mission of Progress-TT: "Scouting ideas/technologies from the PRO and incentivising researchers to disclose IP"; "Assessing IP potential, validating technologies and incentivising for commercialisation"; "Accessing finance and interacting with financial stakeholders"; "Securing TTO Staff skills and organising the TTO for optimum growth".

1 CAF 1: "Scouting ideas/technologies from the PRO and incentivising researchers to disclose IP"

1.1 Copenhagen Spin-outs project at the University of Copenhagen

<u>Critical Area of Focus 1</u>: "Scouting ideas/technologies from the PRO and incentivising researchers to disclose IP"

M. Torrisi (✉)
MITO Technology, Milan, Italy
e-mail: marcello.torrisi@mitotech.eu

© Springer International Publishing AG, part of Springer Nature 2019

M. Granieri and A. Basso (eds.), *Capacity Building in Technology Transfer*,
SxI - Springer for Innovation / SxI - Springer per l'Innovazione 14,
https://doi.org/10.1007/978-3-319-91461-9_9

Aimed at: TTOs/PROs/Policy-makers
University: University of Copenhagen
TTO: Technology Transfer Office at the University of Copenhagen
Country: Denmark
TTO Base: https://ipib.ci.moez.fraunhofer.de/companies/university-of-copenhagen-tto-facility-unknown
TTO year of foundation: 2003
TTO size (nr. personnel): 15
Technology sectors: Medical Technology/Industrial Biotech and Food/Drug Development

The Context

Established in 1479, the University of Copenhagen is one of the largest institutions of research and education in the Nordic countries. The university has over 40,000 students and more than 9000 employees. The university consists of 6 faculties and around 100 departments and research centres. The university has four campuses located in and around Copenhagen, with the headquarters located in central Copenhagen. The Technology Transfer Office (TTO) was established in 2003. The TTO is an internal unit that belongs to the Division of Research and Innovation. The TTO has a team of 15 people, including 7 legal experts and lawyers, 5 business developers, and 3 people generalist team members. All business developers have hybrid scientific and industry background.

The Problem

In 2010, the TTO at the University of Copenhagen faced the following main challenges. First, there was not enough awareness at the level of departments and research centres about the activities conducted by the TTO. Second, there was the necessity to speed up commercialisation activities, in particular the creation of new spinout companies, by engaging more researchers in this arena. Third, a more convinced commitment by the Heads of Department was required to spread the entrepreneurial culture among researchers and facilitate the interactions with the professionals from the TTO.

The Solution

In 2011, the University of Copenhagen together with two other institutions (Technical University of Denmark and the Hospitals of the Copenhagen Region) became part of an EU-funded project called Copenhagen Spin-outs which had a goal of fostering the transformation of life science research into the creation of sustainable spin-outs. The project was (and continues to be) a partnership between the TTOs, industry associations (Danish Biotech, the Pharmaceutical Industries in Denmark), seed capital providers and research parks. Together these actors formed a partner group/steering group for the project. The group meets approx. 4 times a year where spin-out projects are presented, inputs are received and general progress with cases and the overall project is ensured. The set-up is a genuine partnership where all relevant commercialisation stake-holders work together to achieve the common goal of creation more sustainable spinouts from the three research institutions.

In particular, the project aimed at turning existing and new ideas into spin-outs and focused on three main areas: Medical technology, Industrial biotech and food, and Drug development. The Copenhagen Spin-out project added resources that allowed the TTOs at the University of Copenhagen, the Technical University of Denmark and the Capital Region to recruit one-two technology scouts and significantly strengthen the scouting programme. These technology scouts by meeting heads of departments, researchers, and research groups tried to identify and bring out promising ideas residing within departments and research centres. Moreover, technology scouts regularly attended internal research group meetings and idea pitch competitions by researchers and students.

In addition to scouts, the CSO project involved a group of industry mentors. Mentors are experts from companies or other external organisations who can help scouts and researchers on particular issues related to the commercialisation of novel ideas and inventions. Mentors work pro bono. Finally, a series of culture-creating activities were established, such as courses and training for researchers and for university and department management.

A very important aspect of successful technology transfer, in general, and Copenhagen Spin-outs project, in particular, is related to a support and involvement of the university management. For instance, the Pro-rector for Research and Innovation in the majority of speeches at the university highlighted the importance of invention disclosure, patenting, commercialisation, and creation of spin-outs. For what concerns other types of incentives to engage in technology transfer, due to the support of the university top management, a powerful incentive was the inclusion of innovation activities in the annual performance review of departments and researchers, in addition to teaching and publication activities. This was also pushed by the Ministry of Research and Education. Every year each Danish University has to renew a contract with the Ministry, and commercialisation activities are included as performance goals. Moreover, from 2011, the University of Copenhagen and its TTO established an Annual Innovation Prize at the University of Copenhagen, as a way to increase awareness about technology transfer activities and incentivise researchers to engage in technology commercialisation. The prize of 25,000 DKK (3500 EUR) is presented annually to a researcher or a group of researchers, who has made an outstanding effort to create new knowledge and brought it use for the benefit of society. The faculties nominate candidates for the Innovation Prize. Every faculty can nominate a maximum of five candidates. Each faculty decides how to select its candidates. A winner is selected by an international committee (made of internal and external members) among researchers nominated by academic groups. The prize is publicly presented at the university's birthday celebration in November. The Annual Innovation Prize gives researchers public exposure for engaging in innovative activity such as spinning out a new company or successfully licensing an academic invention.

Alignment to Progress-TT

Between 2011 and 2014/15, when the Copenhagen Spin-outs project was undertaken, the institutions created 23 spin-outs, the number of invention disclosures at the University of Copenhagen increased from 40 to 77, the number of patent applications increased from 11 to 24, the overall patent portfolio increased from 58 to 112, and the number of research agreements with companies increased from 306 to 382 per year. Technology scouts were a central component in the new approach to commercialisation and entrepreneurship activities, they were confirmed even after the ending of the CSO project, they also grew in number by adopting the new name of Business Developers. So far the spin-out projects (a total of 23 from 2011–2015) have attracted funding of around 90 million DKK (1.2 million EUR), a clear indication of the quality of these spin-outs.

The success of the project has resulted in the continuation of the collaboration between the partners behind the project (both TTOs, seed funders, research parks, industry associations, etc.) after the EU funding has ceased. A memorandum of understanding was signed in May 2015 where the partners express their intentions to continue to collaborate in the form established during the EU project for another 4 years. The research institutions now fund the scouts out of their own budgets and the scouts are now part of the business development teams of the research institutions' TTOs.

The experience of the University of Copenhagen is instructive to other TTOs for several reasons, but it requires a series of complementary activity and an overall innovation culture to really lead to a steep increase in results. First many researchers have little knowledge about technology transfer and resources available to TTOs; therefore, it is important to organise training events and meeting sessions with researchers and research groups to increase their awareness. Second, the collaboration and support of university top management, including central administration and local management at the department/institute level, is fundamental to legitimise initiatives intended to promote technology transfer and collaboration with industry. It is therefore important to set up meetings with department heads and to be involved in department meetings. Third, incentive systems for researchers, both symbolic and monetary ones, implemented at the university level also play a fundamental role to guarantee that innovation become a part of everyday life at the university. Fourth, it is crucial to create as much publicity as possible around innovation prizes and competitions to increase their effectiveness and involve a larger participation by Faculty members and students. Such competitions can also be combined with additional education and training activities, so to enhance their ultimate impact on technology transfer. Finally, as with other EU-funded (or nationally/regionally funded) projects with limited duration, it is important to create the conditions and the engagement of key-partners in order to assure the continuation and sustainability of the technology transfer initiatives after the project period. It is also necessary to carefully manage the expectations of researchers in this respect, in order to avoid the emergence of distrust and dissatisfaction in case of discontinuation (or partial continuation) of technology transfer support activities.

1.2 Business Development centres at Ghent University

Critical Area of Focus 1: "Scouting ideas/technologies from the PRO and incentivising researchers to disclose IP"

Aimed at: TTOs/PROs
University: Ghent University
TTO: DISCOVERe (Business Development Centre)
Country: Belgium
TTO Base: https://ipib.ci.moez.fraunhofer.de/companies/ugent-techtransfer
TTO year of foundation: 1998
TTO size (nr. personnel): 50 (TTO and BDCs)
Technology sectors: All sectors; Biomedical/Pharmaceutical Research

The Context

Established in 1817, Ghent University is one of the major Belgian universities with over 41,000 students and 9000 employees. The university's 11 faculties are composed of 117 faculty departments. UGent TechTransfer is the central Technology Transfer Office (TTO) of Ghent University. It was established in 1998. The UGent TechTransfer has a team of about 30 people with broad industry experience. In addition to the central TTO, the University established a network of Business Development Centres in 2008 focused on particular technological and application fields.

The Problem

One of the problems encountered by the central TTO was the necessity to deal with a broad range of various technological fields active in different application domains in a generalist University such as UGhent. The challenge of offering/housing in-depth expertise and industry network in a particular field of research from within the TTO was increasingly becoming a problem with respect to scouting promising inventions and favouring their business development activities. The rising awareness of the value of creating IP resulted in a growing amount of patents, but many of these patents remained then unexploited due to the challenge of offering very specific business development advice for such a large number of researchers. Another problem was related to a perceived communication barrier - related to the physical distance - between the "central" office and the "decentralised" scientists and researchers. This perception seemed to sometimes limit the engagement of researchers in the technology transfer process.

The Solution

The solution to the abovementioned challenges was the creation in 2008 of a bivalent/dual model (tandem system), where the TTO was complemented by a set of new Business Development Centres (BDC), so to place scouting and business development activities closer to researchers. Each BDC focused on a particular technological/industrial field. Initially, with the financial support of the Flemish

Government, 4 centres were created, that positively applied to an internal Call of the University (bottom-up approach). A yearly dotation of funding for Flemish universities is dependent on a set of KPI's set out by the Regional Government (in 2015, the dotation amounted to 8.5 million EUR, about 30% of which is used to fund the Business Development Centres). Today, 10 years later, there are 21 BDC in various technological areas (i.e. Energy and Cleantech; Materials; Electronics and Photonics; ICT; Medical; Pharma; Biotech; Food). A particular set of research groups, called clusters, is linked to a certain BDC. Each BDC has a Business Developer who acts as Key Account Manager for the researchers on the one hand, also in their relation with the TTO and for the industry on the other hand. Although the BDC are an internal part of Ghent University, they are not managed by the central TTO and have a high degree of independence. The central TTO acts as an expert service provider for them. The Business Developer supports the researchers of the group in various ways; by identifying opportunities with high commercial-isation potential (internal scouting), by facilitating contacts with industry (external scouting), by helping them with developing their technologies (Proof-of-concepts, tests, etc.) or applying for funding. Once an opportunity is labelled as an innovative project with economical potential, a project team consisting of the Business Developer together with the TTO experts is set up in order to further develop the IP, Legal and Business Development activities in a tandem way. In this way both the BDC as the TTO office interact complementarily together with the scientific experts, being the researchers involved.

The internal scouting activity of Business Developers of BDC can be divided in two types: (1) Reactive/active scouting for valuable opportunities through univer-sity network and resources, matching these with industry insights built during contacts with industry; (2) Pro-active—setting up own strategic research collabo-ration programs focusing on high unmet industry needs, engaging strong research groups with expertise related to such needs. For instance, in the case of DIS-COVERe (the BDC assisting researchers in translating innovative biomedical and pharmaceutical research) the Business Developer undertakes technology develop-ment programs based on opportunities arising from classical research, characterised by bottom-up flow. The Business Developer is either contacted by a group of researchers or contacts them himself to solicit innovative ideas. One method for him to identify projects is to scan through the list of granted basic research projects or titles of Ph.D. theses. Upon identification of ideas with commercialisation potential, the business developer meets the researchers to discuss the project. If a project seems promising it is put into a monitoring list and researchers are regularly con-tacted to receive updates about the progress. The Business Developer manages also his network of industry contacts that can give him some opinions/feedback about the project and its attractiveness for a particular industry/application. Such external scouting activity is also an important component of the Business Developer's tasks, and it is maintained also through regular participation to major Industry Fairs. If the project is found to be promising it gets to the development track. Currently there are 10 Technology development programs ongoing and 5 projects are listed up in the monitoring list of the BDC DISCOVERe. As a part of pro-active technology scouting, DISCOVERe also sets up Strategic research and collaboration programs (often together with industry through R&D collaboration projects) that focus on very risky research projects in which industry is highly interested. These projects are supported in the application for applied research funding from the Proof-of-Concept program financed by the Flemish government and managed internally within the TTO. In the past 7 years, 4 of such Strategic research programs were set up and are currently ongoing within DISCOVERe.

The incentive system for engaging researchers in technology transfer at Ghent University is based on a revenue distribution scheme. The net income (once costs are deducted, such as IP-related costs) out of commercialisation revenues is divided amongst the different stakeholders, being in the first place the department or laboratory from which the invention originates and the inventors, as well as Ghent University itself. Moreover, increasingly indicators of innovation activity (i.e. number of patents applied for, number of spin-offs established, university income via direct industry funding) play an important role at the University and the national level. Although they play a minor role in the career development track of the University scientists, they become relevant when applying for commercialisation/innovation-oriented projects.

Alignment to Progress-TT

The implementation of the above described dual innovation model within Ghent University lead to successful results in terms of scouting and technology transfer activity. Since 2008, when BDC were established at Ghent University, the number of invention disclosures doubled from about 55 to about 110 per year. University patenting activity increased at a similar rate, rising from around 30 patent applications to around 60 patent applications. Furthermore, BDC became more effective in developing partnerships with companies. In particular, 5 out of 10 Technology development programs currently managed by the BDC DISCOVERe have partnering companies that pay for (cofinance) the development costs and have option clauses for the exploitation of the technology in their agreements. In 2008, 4 spin-off companies were established at Ghent University, and 10 spin-off companies are expected to be established in 2015.

The experience of the unique dual innovation model via BDCs and of DISCOVERe in particular is instructive to other TTOs for several reasons. Firstly, it shows how having technology transfer experts with innovation skills close to researchers is constructive to build a climate of trust and collaboration with researchers and to lower the communication thresholds. Secondly, it shows that the specialisation of BDC and their proximity to research groups leads to a clear awareness of moving research results into commercial use and thus successfully stimulating economic development. Thirdly, it suggests the importance of engaging in pro-active scouting activity, trying to match strengths and skills of research groups of the university with technological needs perceived as highly relevant in the industry. Business Developers thus play a role of gatekeepers, and external scouting activities are equally important for them in order to build a network of potential partners and raise awareness on the industry needs.

The experience of Ghent University can probably be more appropriate for large generalist universities with previous history of technology transfer. It may also be adequate and applicable for multi-university or regional technology transfer schemes. In establishing such system with separate business development centres that are independent form the central TTO, it is important to avoid potential frictions between the centres and the TTO. Therefore, it is necessary to organise and structure the system in a way that enhances cooperation and decreases the probability of internal competition.

2   CAF 2: "Assessing IP potential, validating technologies and incentivising for commercialisation"

2.1   Assessing IP and technology at the University of Liege

<u>Critical Area of Focus 2</u>: "Assessing IP potential, validating technologies and incentivising for commercialisation"

<u>Aimed at</u>: TTOs/PROs/Policy makers
<u>University</u>: University of Liege (ULg)
<u>TTO</u>: Interface
<u>Country</u>: Belgium
<u>TTO Base</u>: https://ipib.ci.moez.fraunhofer.de/companies/universite-de-liege-linterface-entreprises-universite
<u>TTO year of foundation</u>: 1989
<u>TTO size (nr. personnel)</u>: 50
<u>Technology sectors</u>: All sectors

The Context

Established in 1817, the University of Liege is a major public university in the French community of Belgium with 22,000 students, and more than 4500 staff members. The university has 11 faculties (colleges) that cover all academic disciplines. The University of Liege ranks among the 10% best universities in the world, considering the World University QS Ranking and among the 20% best universities according to the Academic Ranking of World universities.

The university is located in Wallonia region, the French-speaking part of Belgium. In Belgium, the regional government has prime competencies for science, technology and innovation. The policies for innovation and tech transfer are coordinated through a specialized Agency for Technological Stimulation (Agence de Stimulation Technologique), established by the Walloon government in 2006.

The Interface is a Technology Transfer Office (TTO) of the University of Liege that was established in 1989 to organise and implement the third mission of the university. The Interface was the second TTO created in Belgium after the TTO office of KU Leuven. The Interface is an internal department of the university directly related to the rector's office. For managing technology transfer and proof-of-concept (PoC) funding, the University of Liege has established a commercial company called Gesvel, fully owned and controlled by the university. Both the Interface and Gesvel are directed by the same person and in total have a team of about 50 people with different competencies and extensive industry background.

The Problem

The first problem was associated with the need to develop a PoC funding in Europe, as Belgian universities had a problem of premature technologies that were not yet attractive to industry. The universities had to ask private investors to set up spin-off companies to check and to validate these early stage technologies. However, private investors were reluctant to get involved in such risky projects and were requesting additional validation and de-risking activities. Therefore, in 2007–2008 universities in Wallonia asked their regional government to initiate a policy to set up a PoC funding scheme for universities in the region.

The second problem was associated with selecting the most appropriate projects for PoC funding from 50–60 on-going projects at the University of Liege. Since it may take 2 or more years for maturation of a project, it is important to choose the right moment for entering the PoC funding in order to get new deliverables, which can get a project to the next stage. If a PoC funding is joined too early there is a risk that within the dedicated time frame the milestones may not be achieved and the project may fail.

The Solution

To resolve the first problem, in 2010, the Walloon regional government established a PoC funding scheme, under which each large university in Wallonia received 500,000 EUR per year. At the University of Liege an average budget of PoC fund is about 75,000 EUR per project, but the amount can range from 30,000 to 100,000 EUR per project. The TTO supports about 6 projects per year and the usual duration of PoC funding programme is about 12 months. The TTO is responsible for assuring that researchers follow an accurate action plan and provide timely deliverables. That is partially done by controlling and managing all the expenses associated with the PoC activities by the TTO team.

To resolve the second problem and select appropriate projects for PoC funding, the TTO uses two formalized assessment tools. The first tool is a customized version of the Intellectual Property (IP) score assessment tool developed by Danish national IP administration and made available to other TTOs in Europe by the European Patent Office (EPO). According with this tool, there are two main criteria for selecting projects. The first criterion consists of checking the ownership (i.e. identifying all parties involved in the invention process, including any external parties) and identifying the most appropriate means of protection (i.e. to choose whether to patent or use secrecy). The second criterion involves understanding how far is the technology from the market and what are the next steps necessary to achieve a final product. The closer the technology to the market, the more likely it will be selected.

The second tool is an IRL (Innovation Readiness Level) scale, which is a modified form of a TRL (Technology Readiness Level) scale, which is widely used in different industries, especially in space and aerospace industries to estimate technology maturity. In addition to technical assessment, the IRL scale also includes IP and marketing assessment of a technology. The IRL scale is intended to depict the development of innovation and ranges from IRL-1 to IRL-9. IRL-1 means that a scientific research begins to be translated into applied research and development. Instead, IRL-9 level means that application of the innovation is in its final form and under real-life conditions. The assessment of projects is done in very close cooperation between researchers, the TTO, and external industry people, if necessary. In particular, the PoC programme finances IRL-4 projects to take them to IRL-5. For engineering sector, IRL-5 means that at least a university prototype has been tested in a laboratory. For biotech sector, it means that at least some toxicology tests or some in vivo tests in the actual conditions are performed. The evaluation is mainly made by TTO team, and when it is in a spin off process, external expert (consultants or seed investors) can be involved.

Alignment to Progress-TT

By the end of 2015, the University of Liege totals 26 PoC projects, among which 22 have been completed. This lead to 6 spin-offs created with an average capital of 93,000 EUR at launch. Among those 6 spin-offs, 4 were created between 2010 and 2013 and have today, after new financing rounds, an average capital of 1.6 million EUR. PoC projects valorisation success rate is on average 50%, with a higher success rate for ICT and Engineering projects (85%) than for life sciences projects (33%). On average, the rate of projects financed by the program, as a percentage of the total number of projects which applied for funding is about 30%, i.e. 6 projects out of 20 per year.

By now, the University of Liege has established more than 70 spin-off companies that created about 2000 new jobs. 85% of these spin-off companies are still active. The university has also more than 1000 research contracts with the industry.

The experience of the Interface is instructive to policy makers and other TTOs for several reasons. First, it provides a positive example of how regional governments can initiate policies and PoC funding schemes in order to enhance the transfer of technologies from universities to industries. Second, the example of the PoC programme at the University of Liege highlights the importance of timing and careful selection of projects for PoC funding. It also suggests the importance of adopting systematic assessment tools (based on a qualitative valuation logic) in order to take the decisions on which projects to select into the program. The Interface uses multiple tools to assess the readiness of technology that can also be adopted by other TTOs in Europe. Finally, the case suggests that it is essential to manage PoC funding in a very controlled way, especially when larger amount of money is provided, on average, to selected projects. A process based on the subsequent stages of funding provided to awardees, based on the achievement of interim milestones, can help to deal with risks of underperformance.

2.2 Commercialization Bootcamps and Venture Launch Accelerator at University College Dublin (UCD)

<u>Critical Area of Focus 2</u>: "Assessing IP potential, validating technologies and incentivising for commercialisation"

<u>Aimed at</u>: TTO/Researcher
<u>University</u>: University College Dublin (UCD)
<u>TTO</u>: UCD Innovation
<u>Country</u>: Ireland
<u>TTO Base</u>: https://ipib.ci.moez.fraunhofer.de/companies/university-college-dublin-tto-facility-unknown
<u>TTO year of foundation</u>: 1988
<u>TTO size (nr. personnel)</u>: 20 people
<u>Technology sectors</u>: All

The Context

University College Dublin (UCD) is Ireland's largest university with over 30,000 students and around 1500 academic staff. It has its origins in the Catholic University of Ireland in 1854. The university consists of seven colleges which represent the full spectrum of academic sciences. The technology transfer office of UCD, UCD Innovation that is an internal unit, was founded in 1988 and was a one man show for a decade. In 2002 extra staff was hired for the first time and in 2006/2007 a specific tech-transfer team was established. Nowadays it has more than 20 employees. The TTO is run by top level academy people with also industry expertise: a director who holds a Ph.D. and MBA, 4 Ph.D. level case managers, 3 of whom with MBA, a part time legal counsel and one administrative person. There are also a communication/case Manager with a MBA, a half time person for professional education (masters), an enterprise developer (Ph.D.), a operations manager (for incubation at NovaUCD) plus legal office and finance office within the University. NovaUCD is the hub for new ventures and entrepreneurs at University College Dublin. It is a special office space for UCD Innovation. NovaUCD, which officially opened in October 2003. The concept for NovaUCD was to restore the original house as the centrepiece of a complex of subsidiary buildings that surround it. The buildings are bright, airy and open with high-quality shared and circulation spaces that encourage the formal and informal interactions necessary for the development of a community of entrepreneurs. It also does the research collaborations/IP agreements and has 13 employees, from which 5 to 6 are technology transfer people. UCD Innovation is one of the TTOs run in Europe with employees with very high level competences.

The Problem

Cultivating entrepreneurship is a hard task in a university setting. The gap between academics and business is often large and there are barriers to communication. Without an overarching mission/vision it is impossible to truly create an entrepreneurial culture within the university. Researchers, staff and students might be extremely skilled in their academic research field and in creating radically new inventions and technologies, but they often lack the awareness of the steps leading to successful commercialisation and the required competences.

The Solution

UCD Innovation has implemented several programmes for boosting entrepreneurship and innovation activities. They use a rather personal approach in their everyday work where the case managers work really hand-in-hand with researchers to accelerate the tech-transfer and tech-maturation at the campus. They believe that the team behind the idea is important and try to update themselves regularly about what is "hot" at the campus tech-wise. They actively search and contact key researchers and research projects to offer assistance (for about 55–60 projects in the current year). Scouting ideas at UCD is an internal process where case managers keep an eye on leading researchers and via personal contact look for new entrants for the various programmes. They have implemented an Invention Disclosure Form, based on which the tech-transfer team makes the decision either to patent or not the idea, or to commercialize it or not. Testing the IP from a licensing perspective is done by a team of 4 tech-transfer case managers who know the research on the campus. They focus on ICT, Life Medical Science, Food/Agriculture, Physical Sciences, Mechanical Engeneering/Chemistry.

UCD focuses on fostering the impact of research activities on the wider economy through 2 main programmes: UCD Commercialisation Bootcamp and VentureLaunch Accelerator program. The UCD Commercialisation Bootcamp is a support program for academics and staff at University College Dublin and National College of Art and Design with the aim to provide their researchers, staff and post graduate research students with the knowledge, skills and understanding of the technology commercialisation process and strengthen the pipeline of commercial opportunities arising from UCD and National College of Art and Design (NCAD) research programmes. It takes place twice a year in spring and autumn, and it is delivered by the UCD Innovation team. It consists of five 3-hour practical and hands-on workshops delivered over a 5-week period, where project teams are mentored in-between by UCD Innovation case managers. The submission process consists in filling out an application form, asking for the project name, up to four other project members that attend the Bootcamp, whether the project has an invention disclosure with the UCD TTO, a short description of the market problem, the current stage of development and the current funding of the project. UCD chooses around 10 participants for the Bootcamp per year. On completion of the Bootcamp participants will have developed a solid awareness of what it takes to build a commercialisation plan which covers the following areas:

1. Market problem/need
2. Proposed solution/technological innovation
3. Market opportunity and market validation
4. Commercial exploitation route(s)

5. Project team
6. Technology roadmap
7. Commercialisation work plan
8. Funding requirements and funding plan.

Attendants will also create a commercialisation canvas, based on the Business Model Canvas framework, and it incorporates Design Thinking and Lean Start-up techniques and helps them to map out the most optimal commercialisation route. Through such techniques participants are encouraged to obtain an early and fast feedback from potential users on their ideas, so to identify promising areas for development and abandon less interesting avenues. The method will generate a so-called "living document", which is not a static structure for the commercialisation but gives room for pivoting and updates. The idea is to actively and dynamically develop a commercialisation route, especially using user feedback and an experimentation-driven approach.

UCD Innovation also runs the VentureLaunch Accelerator program, a complementary programme with the aim to support the creation of new ventures based on UCD intellectual property. Ten three-hour workshops over three months will help the researchers acquiring the knowledge, skills and understanding that will be required to develop teams and technologies into commercial ventures. At the end of the year, a winner is chosen who will get 10,000 Euros of seed money plus 15,000 Euros usable for professional services, as a way to foster the grow of the new venture. Companies supported by Commercialisation Bootcamp and VentureLaunch programme in the last 10 years have raised over 100 million Euros in investments, have created around 1600 new jobs. Since 2008, there have been 4 trade sales as exits, realization for them is around 170 million Euros, 3 big trade sales with 150 million Euros and a smaller one of 6 million Euros (time period from 2003 until 2013).

Various funding streams are used for implementing the programmes. The tech-transfer group in UCD is funded by the Innovation Agency Enterprise Ireland (Tech Transfer Strengthening Initiative), although being employed by UCD. That means that the people working for NovaUCD are employed by the University, but the funding for their salaries comes from Enterprise Ireland. In particular, Enterprise Ireland provides funds for commercialisation, helps getting a consultant with great knowledge in market/tech area for commercialisation for projects provides funding for market research: feasibility grant, that will help getting commercialisation funds. Another source of funding is Science Foundation Ireland (TIDA). Finally, there are the following extra funding schemes from NovaUCD: company/industry and university research Innovation Partnership, Irish Research Council, Enterprise Partner Scheme, Employment Based Ph.D. Scheme (it pays leave of absence for business people to undertake a research Ph.D.). There are specific milestones for each sort of funding process within UCD Innovation, but the first milestone is common to all the programmes and consists in the management team looking at the idea and giving it a "go" or "no go". That in many cases is made by the managing director himself. UCD Innovation is really focused on helping the researchers getting external funding. At UCD Innovation NovaUCD applies a close collaboration approach where case managers give advice to researchers and help them writing application forms. The case manager and researched working together also often help to secure grants/funding for research.

Several metrics have been taken into use at UCD Innovation to evaluate performance, partly due to the reason that funding from Enterprise Ireland requires that. They report quarterly on invention disclosures, patent filings, license options and agreements, spin outs, engagements/agreements with industry. They benchmark their numbers against numbers they find internationally in the US/UK and other countries, for example they look for 2 ideas every 1 million euro of research income in the university, 1 patent application per 5 million euro of research income and 1 license agreement for 10 million euro research income. UCD has approximately 55–60 invention disclosures per year. They file around 20 priority patent applications, make on average around 20 licensing agreements and 3–4 spin-offs per year. UCD research income staggered a bit over (about 50% less) the last years due to the negative effect of the crisis. Nevertheless, the quality of UCD Innovations is more important than the volume.

Alignment to Progress-TT

In 2013, 37 and 17 researchers attended the first two UCD Commercialisation Bootcamps, representing a total of 30 potential commercial projects. The third one was completed by 24 researchers in 2014. Overall, 270 companies and early-stage projects where supported at NovaUCD (the incubator of UCD where UCDInnovation and the Bootcamps take place), 140 companies where incubated and 30 new UCD spin-out companies incorporated. The UCD Commercialisation Bootcamp is part of a larger initiative to foster the entrepreneurial spirit of researchers. First, they can attend "sprints", one-day activities that help commercialisation, going deeper with boot camps and then use the VentureLaunch program which should ultimately lead to new companies created. This process streamlines and refines the commercialisation process that the researchers started. Using the well-known Business Canvas Model and improving it for the specific demands of university staff that want to commercialize their research, providing a flexible route and utilizing user feedback, the boot camp at UCD is a good example of experimental, hands-on approach for validating and maturing technologies. An additional lesson that can be learned from the experience of UCD Innovation is that a TTO is only one part of the innovation ecosystem. An active industrial sector, with both large companies and small-medium enterprise (SME) which can absorb/license technology is crucial as well. Ideas need licensees, investors are needed in both the research and the SME side.

3 CAF 3: "Accessing finance and interacting with financial stakeholders"

3.1 The partnership with IP Group at Cardiff University

Critical Area of Focus 3: "Accessing finance and interacting with financial stakeholders"

Aimed at: TTOs
University: University of Cardiff
TTO: TTO Cardiff University
Country: Wales, United Kingdom
TTO Base: http://sites.cardiff.ac.uk/bls/the-technology-transfer-team/
TTO year of foundation: 1990
TTO size (nr. personnel): 7
Technology sectors: Arts, Humanities and Social Sciences/Biomedical and Life Sciences/Physical Sciences and Engineering

The Context

Cardiff University is a public research university located in Cardiff (Wales, United Kingdom). It is the third oldest university institution in Wales and is the only Welsh member of the exclusive Russell Group of leading British research universities. Cardiff University is composed of three colleges: Arts, Humanities and Social Sciences; Biomedical and Life Sciences; and Physical Sciences and Engineering. It employs more than 6000 staff and has around 30,180 students (it is the 8th largest university in the UK in terms of student numbers).

Its TTO was established in 1990. It is an internal unit of Cardiff University and is a part of Professional Services. The staff has a team of 7 people. All the staff members are employed by the University. The staff is divided among the three Colleges that compose the University according to their size: 3 staff members deal with the College of Biomedical and Life Sciences; 2 with the College of Physical Sciences and Engineering; 1 with the College of Arts, Humanities and Social Sciences and another one covers all of them. The TTO covers all the technology sectors of Cardiff University but most of its activities are related to the College of Biomedical and Life Sciences. The backgrounds of the people employed at the TTO are quite broad and cover all the commercial, research and intellectual property aspects.

The activities of the TTO are principally focused on developing innovative research by securing translational funding. Working alongside academics, the TTO advises on the completion of grant applications and it can provide the required intellectual property statements specific to a project. It then further develops these research outcomes by protecting the resulting intellectual property and by exploiting its specific know-how through negotiation of license deals, working with industrial partners and the creation of spin-outs.

The Problem

One of the main challenges faced by the TTO was to organize in an optimal way the partnership between Cardiff University and VC firms. In particular, there was the need to fill the gap between Cardiff University's TTO and traditional venture capital investing, especially considering that the TTO is principally focused on Biomedical and Life Sciences activities and its technologies are often early-stage. In addition, there was the need to effectively manage the early incubation stages of spin-out companies.

The Solution

University of Cardiff addressed the first challenge by partnering with an outside Investor group, Fusion IP plc (now part of IP Group plc) with whom it had a long standing relationship. Fusion IP plc, previously called Biofusion, was established in 2002 to commercialise university-generated intellectual property. In 2014 it was acquired by IP Group (which has owned a 19.8% stake in Fusion IP since 2009), in order to create a stronger UK-based IP commercialisation company. IP Group (a publicly-listed intellectual property company) works on commercialising IP sourced from research institutions and it pioneered the concept of the long-term partnership model with UK universities. It now has arrangements covering fourteen of the UK's leading universities and also has access to opportunities in the Oxford and Cambridge clusters from its stakes in Oxford Sciences Innovation plc and Cambridge Innovation Capital plc respectively. Its portfolio comprises around 100 companies and is categorised into 4 main sectors (Biotech, Cleantech, Healthcare and Technology).

The partnership between Cardiff University and Fusion IP/IP Group is one of the most important features of the TTO and it has proven to be very successful for Cardiff University. It was completed in 2007 and it consists in a ten-year exclusive agreement with Fusion IP (which was acquired successively by IP Group), giving it the right to commercialise, through the creation of spin-out companies, Cardiff University's research-generated intellectual property. This agreement was very similar to that which the same company completed with Sheffield University. Currently, IP Group has 7 companies in partnership with University of Cardiff: Alesi Surgical Ltd, Fault Current Ltd, I2L Research Ltd, MedaPhor Group plc, Morvus Limited, Nanotether Ltd and Progenteq.

The original terms and arrangements of the agreement between Fusion IP and Cardiff University has not changed after the acquisition of the company by IP Group. According to the terms of this agreement, IP Group has the first right to establish a spinout company based on Cardiff University owned intellectual property. In return for this exclusive option, Fusion IP gave the University a significant shareholding in Biofusion (30%). Furthermore, it provided a ring-fenced fund of £8.2 million for investing in spinout commercialisation opportunities arising from Cardiff University's research. If Fusion decides not to invest in a specific Cardiff University's Spin-out within 30 months, the TTO has the total freedom to pursue all the alternative strategies and means that it considers appropriate for the exploitation of the related intellectual property.

The advantage of this type of partnership consists in having a professional investor who raises funds for the University and it is willing to invest earlier than traditional venture investors, when there is an important technology risk.

Cardiff University addressed the second challenge, i.e. creation of successful spin-outs, by means of the same relationship, leveraging the management and business support that Fusion IP provides. This is very important, especially considering that the Cardiff University's TTO is principally focused on Biomedical and Life Sciences activities and its technologies are often early-stage.

The creation of a spin-out is structured in 3 stages: Start up, later stage investment and trading. The first step usually lasts 12/18 months. In this step Fusion IP is very involved: it provides support in writing business plans, it builds the website and it deals with early-stage company formation issues. IP group usually funds this step on its own. In the second step, which usually lasts 12/36 months, Fusion IP provides support in developing the business of the company including drafting the business plan, recruiting the management team and in raising third party funding. Finally, the last step is characterized by the attempt to reach a trade sale or by IPO (Initial Public Offering).

As far as the partition of equity is concerned, according to the terms of the agreement, Fusion IP normally owns the 60% of any new portfolio company at start-up and the remaining 40% goes to the academic funding group. Cardiff University does not take an equity stake in individual spinouts, but is a major shareholder in Fusion IP plc. This has the advantage of aligning the goals of the investors, the executive management of Fusion IP and the University. It also helps mitigate the University from any downstream dilution in individual spin-out companies.

IP Group's acquisition of Fusion IP in 2014, resulted in a larger entity with greater investment potential than Fusion IP alone. There was also much greater liquidity in the IP Group stock making it possible for Cardiff University to sell its shares in IP Group.

Alignment to Progress-TT

Managing the relationship with VC investors, especially for Biomedical and Life Sciences, is often a difficult task, in particular for the significant investments that the technologies in these areas require.

The partnership that Cardiff University established with Fusion IP (acquired subsequently by IP Group) has proven to be very successful for its spin-out business. In total, Fusion IP/IP Group have invested in 20 new companies from Cardiff University and have raised almost £50 M. Considering that Cardiff University raised additional investment finance to that which Fusion IP/IP Group invested and secured, there were raised more than £75 M in venture funding in total. 4 spin-out companies from Cardiff University have listed on AIM (Alternative Investment Market; it is a sub-market of the London Stock Exchange).

An example of the success of the deal between University of Cardiff and Fusion IP/IP Group is represented by MedaPhor Group plc. MedaPhor is a global provider of advanced ultrasound education and training for medical professionals which was created in 2004 by a team of clinicians and computer scientists. Since 2007 MedaPhor has raised almost £10 M in investment funding from Fusion IP, Finance Wales, IP Group, Arthurian Life Sciences Fund, Cardiff Capital and founding directors. With headquarters in Cardiff (UK) and San Diego (CA, USA), the company has sold over 120 ultrasound systems into major hospitals worldwide, with sales rising to £1.8 M in 2014. MedaPhor became the third IP Group portfolio company to float on the AIM market of the London Stock Exchange in 2014, raising £4.7 M from institutional investors for its flotation. More recently Meda-phor's innovative ultrasound training simulator has been endorsed by the American Board of Obstetrics and Gynecology (ABOG). This means the ScanTrainer will be used in 2000 obstetrics and gynaecology certification exams carried out across the United States each year. The three-year endorsement led MedaPhor to raise £3.2 M through a share issue on the London Stock Exchange Alternative Investment Market. Over 200 hospitals in the US run obstetrics and gynaecology residency programmes.

In conclusion, Cardiff University has established a business model for spin-out formation based on a partnership agreement with specialized investors. This business model is based on a long-term exclusive agreement with an investment firm (Fusion IP/IP Group) specialized in commercialising IP sourced from research institutions. In this way, it is possible to fund the earlier stages of the commercialisation process (an area where traditional venture investors are often not willing to invest, due to high levels of risk) and, consequently, fill a gap between Cardiff University's TTO and traditional venture capital investing. In addition, the partnership with Fusion IP/IP group was very important in order to provide hands-on support and advice for the initial maturation of the early stage spin-out companies. In fact, thanks to their experience, Fusion IP and IP Group were very useful in these particular steps, thus allowing to optimize the growth strategies of spin-out companies in their target markets. They are able to undertake many of the issues associated with the creation of a new company, including, for example, undertaking due diligence and market research pre-incorporation. Where it is decided to set up a company, IP Group can also take care of this activity, including establishing legal agreements and registering the company. Furthermore, it can invest additional sums to recruit management with the required expertise. By doing this, when another investor is approached, the venture is more mature and represents a more attractive investment proposition.

3.2   NIK Seed Funds at Bergen University

<u>Critical area of focus 3</u>: "Accessing finance and interacting with financial stakeholders"

<u>Aimed at</u>: TTOs
<u>University</u>: University of Bergen, Haukeland University Hospital, the Institute of Marine Research, Bergen University College and Siva.
<u>TTO</u>: Bergen Teknologioverføring AS (BTO)
<u>Country</u>: Norway
<u>TTO Base</u>: https://ipib.ci.moez.fraunhofer.de/companies/bto
<u>TTO year of foundation</u>: 2004
<u>TTO size (nr. personnel)</u>: 30
<u>Technology sectors</u>: Oil and gas, aquaculture, marine technologies and medicine

The Context

Founded in 2004, Bergen Teknologioverføring AS (BTO) is a regional TTO serving 10 universities and research institutes in the Bergen area. It is owned by five of them (University of Bergen, Haukeland University Hospital, the Institute of Marine Research, SIVA—Industrial Development Corporation of Norway, and Bergen University College). Together, BTO's partners and owners employ about 4000 researchers with a combined research budget of over EUR 425 million.

The BTO team consists of 30 highly skilled and experienced people covering the full range of technology transfer services from securing IP through to its commercialization. A particular focus is on business development and start-up support. Since 2013, BTO also runs an incubator facility for start-up companies in Bergen and is responsible for the administration of clinical trials at Haukeland University Hospital.

The Problem

In order to move early, high-potential projects effectively towards an advanced technology readiness level, BTO has adopted a proactive approach to start-up development. TTO staff takes on operational tasks and sometimes acts as interim CEO. While this works extremely well during the early phases of project and company development, it becomes increasingly difficult when start-ups grow successfully and funding requirements rise steeply. The amount of time needed to attract appropriate investments goes far beyond what BTO can do with its own resources and funding options.

The Solution

BTO, two further Norwegian TTOs (Kjeller Innovation, Oslo, and NTNU, Trondheim) and Televenture, a leading early phase venture capital firm in Norway, have set-up a new investment company, Norsk Innovasjonskapital AS (NIK), in August 2010. Managed by Televenture, it pursues an innovative funding concept: The fund first selects investment opportunities, i.e. early stage start-up companies from the TTOs portfolios, and then raises money from private investors to develop these companies to the next value level.

It has presently 4 funds under management, NIK I-IV. The first three of them were established as separate funds, one per TTO (mainly because the TTOs entered one after the other into the negotiation process), and the fourth one is a joint fund for all three TTOs.

In total, 28 companies were selected for the funds NIK I-III, and a total of approximately EUR 150 million was raised from private investors for all three funds. There was no contribution from government or public bodies and no cash contribution from the universities and TTOs (but they contributed their respective shares in the companies which were exchanged for shares in the NIK funds). The approach works as follows:

Each TTO develops its start-up companies through the initial stages of project validation, incorporation and early growth. NIK management then screens the TTOs' portfolios and picks the most promising start-ups for the fund. They are looking for innovative, validated technologies and convincing business concepts, but these can be fairly early in terms of their development stage. They cover a broad range of sectors and various business models, including service and product businesses. BTOs experience is that about 20% of its start-ups will be selected by NIK.

The companies of choice are then transferred to NIK in return for shares in the fund. From then on, BTO will not own sharers directly in the startups. After this transaction, the fund managers will market the selected companies to potential investors who are invited to meet them and learn about their projects and projections.

The fund management is responsible for fundraising and developing its portfolio companies to the next value level. The managers themselves contribute to strategic issues on board level and frequently bring in new people at the operational level to lead and support the ventures. The further role of BTO varies case by case. Finally, the fund management will sell either individual companies or the whole fund. NIK I-III comprising a total of 28 start-ups are now to be sold as a whole package, some five years after their initiation.

Alignment to Progress-TT

The approach proved to be extremely successful. Since NIK was established in 2011, it has raised EUR 60 million from 130 investors. About 25% of them have not only invested into the fund, but additionally into individual start-ups from the NIK portfolio. NIK funds have thus attracted more money than has been invested in Norway through seed financing schemes which include risk-reducing matching funds from government. This is remarkable, even more as the start-ups selected by NIK are early stage, i.e. high-risk. The current fund is valued in the three-digit range, as assessed by external experts. Having started out with companies valued somewhere near zero, this represents a dramatic increase in value for the TTOs, their PROs and inventors.

Two aspects clearly differentiate NIK from other early-stage investment tools: First, marketing is based on distinctive high-tech stories opening a world full of opportunities to potential investors. Secondly, NIK management provides a professional, trusted set-up for adding value to these stories. This is a unique combination that obviously sells. The principle can be replicated by other TTOs all over Europe. Moreover, it can be started fairly small and extended to multiple organisations later. It is also possible to pool projects from various sources according to themes and adopt a portfolio approach to development.

A lesson learnt is that a regional focus is not relevant to investors. NIK II was established as a regional fund for the Bergen area, but it turned out that investors came from all over Norway and abroad—just as for NIK I and III.

Critical success factors include significant business development skills together with an entrepreneurial mindset enabling the TTOs to build up an attractive pipeline of projects for the fund to choose from.

Overall, the NIK approach has proven very effective in moving early-stage projects to the next value level while taking away significant workload from the TTOs as the investment requirements of their start-ups rise steeply.

4 CAF 4: "Securing TTO Staff skills and organising the TTO for optimum growth"

4.1 Organising a joint TTO at the Universities of Basel, Bern and Zurich

Critical Area of Focus 4: "Securing TTO Staff skills and organising the TTO for optimum growth"

Aimed at: TTOs
University: University of Basel, University of Bern, and University of Zurich
TTO: Unitectra
Country: Switzerland
TTO Base: https://ipib.ci.moez.fraunhofer.de/companies/unitectra
TTO year of foundation: 1996
TTO size (nr. personnel): 14
Technology sectors: All sectors/Physical Sciences/Biomedical

The Context

Unitectra is a joint Technology Transfer Office (TTO) of the Universities of Basel, Bern, and Zurich originally established in 1996 under the name of Biotectra in the scope of the Priority Program Biotechnology of the Swiss National Science Foundation. The University of Basel, established in 1460, is the oldest university in Switzerland and has around 13,000 students, nearly 400 professors, 7 faculties, and 18 departments. The University of Bern, established in 1834, is one of mid-range size universities in Switzerland and has nearly 17,000 students, more than 400 professors, 8 faculties, some 160 institutes, and 8 graduate schools. The University of Zurich, established in 1833, is the largest university in Switzerland and has around 26,000 students, more than 600 professors, 7 faculties covering some 100 different subject areas.

Based on the work of Biotectra, the Universities of Bern and Zurich established a joint TTO called Unitectra in 1999. As a non-for-profit incorporated company, Unitectra was completely and equally owned by the Universities of Bern and Zurich. In 2013, the University of Basel became the third shareholder of the company after Unitectra had taken over its technology transfer activities in 2011. Unitectra's technology transfer services are also available to researchers positioned at hospitals and research institutions that are associated with the three universities. Unitectra has a team of 14 people, including 7 technology transfer managers, 4 contract managers with a legal background, and 3 finance and administration managers.

The Problem

One of the main challenges faced by Unitectra was to establish a system based on the collaboration between several universities which was able to be very productive despite having a relatively small size.

In addition, by establishing a joint TTO, it was important to avoid any conflict among the universities regarding the management of their technology transfer activities. Another challenge was to manage the cases according to the guidelines of each particular university.

Since 1999 Unitectra is acting as a joint TTO for the Universities of Bern and Zurich and since 2011 also for the University of Basel. The basic idea of the Unitectra founders was indeed to spur universities and their researchers to do more in the field of technology transfer through a collaboration with other universities instead of each institution establishing a subcritical unit on their own not competent to cover all the different areas of research. Considering this, they proposed and convinced these universities to establish a system in which each university funded its own technology transfer activities, without subsidising operations at the other institution, while at the same time taking advantage from the synergies arising from such a collaboration. Originally, Unitectra was financed by the two universities of Bern and Zurich according to a fixed share based on the relative size of the institutions. Nowadays, the three universities finance their joint TTO according to the amount of work done for each institution.

The Solution

Although the three universities are different in terms of size, structure and resources, Unitectra was able to create an organization which has worked for more than 15 years without any problems between these universities. A key point of the success of Unitectra is its high efficiency despite the relatively small size of the TTO, especially considering that it deals with three universities and also with the hospitals and research institutions that are associated with these universities.

Unitectra maintains an office at each of the universities and the team is spread among these three locations. This ensures that researchers at each university have direct and equal access to Unitectra's services. The technology transfer managers flexibly work at two or some of them at all three locations to make sure that they can meet the researchers at the respective location. Thus, Unitectra acts as one team with part of the team rotating among the different offices. This set-up is also important to ensure common quality standards and a joint entrepreneurial culture within the TTO. Another challenge is represented by the fact that the three universities partly have different regulations and Unitectra has to manage the cases according to the guidelines of each particular university. Unitectra succeeded to overcome these difficulties and to reach high standard of performances by exalting the advantages derived from this collaboration and, at the same time, by managing the potential disadvantages. A potential disadvantage of a joint TTO could be represented by the problems deriving from the difficulties in dealing with its universities in an "unbiased" manner. Before its establishment, Unitectra together with the universities carefully tried to identify all the possible issues that could arise and took appropriate measures. Important aspects are a high degree of transparency, regular reporting and the functional integration of Unitectra within each university. This resulted in a situation where each of the three universities fully considers Unitectra as its own TTO.

Through its services, Unitectra is responsible for the commercialization of academic research results (patents, licenses, creation of spin-off companies) and the negotiation of research contracts between the universities and economic partners. It plays a key role to support researchers in the research collaborations with private industry and other public or private research institutions. About 75% of Unitectra's activities are in the field of Life Sciences (Pharma, Biotech, medical devices), the rest in the fields of physical sciences, ICT and the Humanities. This reflects the fact, that the three universities don't have an engineering school.

One of the most important goals for Unitectra is customer focus. The researchers, which are its customers, are seen as an important resource, a full part of the process and so they are involved as much as possible. This is an important aspect that differentiates Unitectra from many other places, where TTOs just take over projects from the researchers and run off on their own. The evaluation of an invention disclosure is made in a very open manner and, at the end of the process, researchers accept and understand the responses without any problem, even if their inventions are not considered worthy to follow up. Researchers are also involved in the technology marketing process and are helped in the communication with companies because in Unitectra's experience it is often better for a company if an opportunity comes from a researcher rather than the TTO. Unitectra takes advantage of the collaboration with researchers in order to increase the possibilities to commercialize research results through their contacts. Thus, researchers are an important resource in the commercialization strategy. The TTO is organized in order to build a trustful relationship with researchers and university management. By having its offices at each university, Unitectra is capable to be very close to them, simplifying interaction and information exchange. In this way, it is possible to have a structure which is considered by the researchers as an internal unit and which is able to develop a reliable working relationship with the researchers of each university. At the same time, this organization allows a sufficient amount of freedom in different aspects because, although the universities are the shareholders of Unitectra, it can operate like a normal company with the required freedom.

Another important key factor is the quality of personnel. Unitectra recruits prospective employees from the private sector or with experience in the private sector: technology transfer manager candidates must have a scientific background and also experience from the private industry. Although most of the employees have a Ph.d., it is not an absolute requirement. More important is a significant amount of experience (at least 5/6 years) in the private sector in a customer-oriented role.

The competences are built principally with on-the-job training with mentors: when one starts to work in the TTO, he or she works very closely with a mentor. This on-the-job-training is completed by internal or external training courses. These courses are selected depending on the needs of the staff. The training process for a technology transfer manager typically lasts about two years, until a person is familiar with the key aspects of the job. Unlike many other TTOs, the turnover of the personnel is very low and Unitectra makes a lot of effort to retain good personnel because a trustful relationship with researchers is considered a key aspect for TTO success, something which needs time to be developed. In fact, half of the team has been with Unitectra for more than 10 years. This is possible by providing a stimulating environment in the office with flat hierarchies and by empowering people as much as possible. Moreover, the compensation needs to be adequate to people's experience, responsibility and quality of work, and reasonably competitive in comparison to the private industry.

A sophisticated case management software that was developed and is maintained by Unitectra in collaboration with an IT company serves as a key resource to properly manage the high number of cases (>2000 new cases in 2015 alone) for the different institutions. This software also is an important management tool providing important data about the performance and productivity of the individual team members and of the organization overall. In addition, it easily allows to extract the necessary data for reporting purposes to university management and for addressing any requests that might arise from researchers, companies or other stakeholders.

Unitectra pays a lot of attention to regular interactions with the management of the three universities and the hospitals. This is facilitated by the fact that two members of the university management of each university are sitting on the board of directors of Unitectra along with industry representatives.

Alignment to Progress-TT

The organization of the TTO has proven to be very successful for Unitectra. Between 2011 and 2015, 70 new priority patent applications were filed on average per year, the number of licensing agreements of the three universities increased from 48 to 67 per year. The number of spin-off companies increased from 8 to 14 per year, research contracts increased from about 1050 to more than 1200 per year, other agreements (MTAs, consulting, etc.) increased from about 600 to about 1100 per year. Under a license of any of the three universities more than 90 products were launched on the market by companies since 1999 and, in the Pharma/Biotech field, more than 30 new molecular entities (NME) are currently in various phases of clinical development by licensees, including spin-off companies.

The experience of Unitectra can be instructive to other universities and TTOs for several reasons. First, Unitectra is a good example of a TTO that is able to deal with three universities and their associated hospitals and research institutions. The synergy between these entities allows to have sufficient volume of projects and to build a TTO capable of covering all the different areas in a professional manner, an important aspect for small TTOs with limited financial and human resources. This model is based on a high degree of transparency towards each university. The difficulties arising from managing a team that is located in multiple places are actually mitigated by the relatively short distances between the three cities (one hour by public transportation). However, it is interesting to point out how, despite the small size, Unitectra succeeded in creating a system which has achieved great results in technology transfer without having any problems in the management of the collaboration between the universities. Another important aspect of Unitectra is the low turnover of the staff: half of the team has been with Unitectra for more than 10 years. Considering this, the model seems to work also in terms of personnel satisfaction.

Second, Unitectra has established a customer focused model based on trust and long term relationship with researchers and university managements that can be replicated by other TTOs in Europe. The close interaction with the researchers and the university management, the focus in the TTO activities, and the experience of its personnel are key success factors. In this way, it was possible to create an organization which takes advantage of the collaboration of the three universities in the field of technology transfer and, at the same time, maintains enough flexibility to prosper as a flexible and independent unit for the benefit of the universities.

4.2   Organise to take advantage of opportunities at IOCB

Critical area of focus 4: "Securing TTO staff skills and organising the TTO for optimum growth"

Aimed at: TTOs
University: Institute of Organic Chemistry and Biochemistry AS CR, v.v.i.
TTO: Technology Transfer Office
Country: Czech Republic
TTO   Base: https://ipib.ci.moez.fraunhofer.de/companies/institute-of-organic-chemistry-and-biochemistry
TTO year of foundation: 2009
TTO size (nr. personnel): 6
Technology sectors: Chemistry

The Context

The Institute of Organic Chemistry and Biochemistry (IOCB) Academy of Sciences of the Czech Republic in Prague carries out fundamental research in organic chemistry, biochemistry and related disciplines, focusing in particular on medical and environmental applications in nucleic acid components, proteins, peptides, natural products, synthetic functional molecules and molecular modelling. IOCB has a renowned area of expertise in antivirotics and peptide drugs and the aim of the Institute is to reach excellence in the international competition and to keep this position in the long term. IOCB provides scientific assessments, professional opinions and recommendations, consulting and advisory services, and promotes science popularisation. In cooperation with universities, the IOCB carries out doctoral study programmes and provides training for young scientists. It is currently home to 600 researchers. IOCB started a collaboration the American biopharmaceutical company, Gilead Sciences and in 2006 both partners established a joint research centre. This collaboration has produced, among others, the HIV therapy blockbuster Tenofovir^{TM}.

The IOCB Technology Transfer Office s.r.o (IOCB TTO) is a wholly-owned subsidiary company of IOCB and a leading technology transfer office in the Czech Republic in the field of pharmaceutical, chemical and biotechnology research and development. They match the research outputs of Czech scientists with the needs of commercial partners to bring new ideas in medicinal chemistry, material sciences, biology and other fields of chemistry to human use or technology market. IOCB TTO currently employs 6 people. CEO; Patent Attorney; General manager; Administration Assistant and two Innovation Managers.

The Problem

IOCB TTO was formed in 2009 with a staff of two persons: a CEO highly experienced in technology transfer and a patent attorney. The model was to offer advice only on patenting and commercialisation issues to the Director of IOCB. There were no resources available for project development or project management. Despite good advice being available, the research teams in the IOCB did not have the skills or resources to develop research outputs to the point of market- or investment-ready products. Consequently, IOCB TTO was not having the expected impact of increasing the output of commercialised research for the parent institute.

The Solution

IOCB TTO realised that a more systematic project management activity was needed to progress ideas through to products and grow the TTO and IP portfolio. Their first steps were to put a process in place to manage the IP portfolio and to start an outreach program to advertise the presence of the TTO throughout the IOCB. There was little extra resource available to fund this activity but a solution to this was found when IOCB TTO successfully bid for financial support from the Technology Agency of the Czech Republic to form a Centre for Development of Original Drugs. This is a collaboration between three private companies and six academic institutes (including IOCB) which is managed by IOCB TTO. The grant came along at a very opportune time, when IOCB TTO were poised and ready to expand their operations. The grant runs for eight years from 1 April 2012 and has provided a stable base in project development funding which has helped to kick-start the expansion of the TTO. The funding for the centre is 65% of actual costs, up to a limit of 40 M CZK/year (€1.48 M) with an evaluation and audit after 4 years to secure the final 4 years of funding. Partners contributions are in-kind and varies from partner to partner. IOCB in-kind contribution is 50%.

A part-time assistant was hired to cover administration of the centre, keeping this from adding to the duties of the TTO staff. Then a general manager was hired to share the workload of the CEO, liaise with inventors and begin to manage the screening and development of research outputs. Since then two further project managers have been hired as the workload of the TTO increased. The first of these is an experienced R&D manager from a pharma company who needed very little further training in technology transfer. The second project manager is a qualified scientist who has research experience in the relevant fields and who is now training on the job.

IOCB TTO currently handles about 15 projects at a time with 5–6 disclosures each year feeding the project pipeline. Each project manager (including the general manager) handles 3–5 projects, according to their complexity and needs. They use external consultants for advice when a project lies outside of their core competencies in bio- or medicinal chemistry. IOCB TTO also enters into collaborations with other institutes for proof of concept or when they need to licence IP belonging to another institute to develop a new product. Although each project manager delivers a project from end-to-end of the technology transfer process, the decision to move a project forward is taken by the team as a whole.

Seed funding from the Centre for Development of Original Drugs is currently funding 9 individual drug development projects shared and developed by the 9 partners. Projects taken up by the centre need an approval of the Technology agency and must have a good business/development plan. Seed funding is allocated according to the needs and priorities of individual projects and is usually between CZK 1 M and CZK 4 M (€37 K–€147 K).

Staff training is a mixture of on-the-job training from more experienced staff and training courses offered by ASTP-proton. One staff member is currently working towards the Registered Technology Transfer Professional qualification (RTTP) through membership of ASTP-proton. This will be offered to all junior and new members of staff in the future.

Funding of the TTO comes from IOCB together with some royalty income. The aim of IOCB TTO is to develop their portfolio and resources to achieve a higher level of commission on royalties from commercialisation deals negotiated by them in the future. This will produce more funds for future project development and commercialisation and enable the TTO to continue to grow.

Currently, the relative low number and high diversity of projects means that there is no one-size-fits-all process to evaluate and manage the projects. Potential products are benchmarked against existing products, revenues and outcomes and this provides an initial valuation. IOCB TTO are developing tools to evaluate projects but at the moment it is better to be flexible in their approach while using the experience gained to improve the processes and develop the tools. The capacity and expertise needed to commercialise research is changing constantly so they are encouraging their staff to develop their skills to do things more effectively in the operation of the TTO and this has resulted in a high rate of growth over the last two years and will lead to bigger deals in the next two. IOCB have 4 current projects which are being prepared for spin-out in the near future. Two projects: iBodies (synthetic polymer conjugates capable of replacing antibodies in biomedical applications) and DIANA have been selected for oral presentation at this year's BioVaria conference May 17th 2016 in Munich, Germany. The Lipeon project (a novel molecule to treat diabetes Type 2 and obesity) was selected for the Spin-off panel at BioVaria. The SpinPanel ges potential start-ups an opportunity to convince top life-science investors and biopharmaceutical decision-makers of their business concept.

Alignment to Progress-TT

This is a great example of how a smaller TTO can organise to take advantage of opportunities and deploy limited resources to deal with a diverse portfolio of projects. IOCB TTO were prepared to move immediately when funding was secured to kick-start growth. They have started by hiring some highly experienced staff but this is not a sustainable practice where resources are limited. As their portfolio is growing they are hiring staff with the required scientific background and increasing their TT knowledge with a mixture of external and on-the-job training. They have a plan to increase their portfolio of commercialised IP and will keep staff capability and numbers in step with this through continuous improvement with training and experience. Working towards the Registered Technology Transfer Professional qualification (RTTP) through membership of ASTP-Proton provides an easily accessible route to external training and certification. Where needed, their knowledge and experience is supplemented by using external experts or collaborating with other institutions for mutual gain.

Critical success factor is that, for successful technology transfer to take place, there has to be a desire from the PRO management to make TT an important function of the institute and with this desire to provide at least some basic funding. Don't expect results overnight as it takes time to develop the resources and experience needed to discover, develop and commercialise research outputs. Fortune favours the prepared mind and smaller TTOs should be ready to take full advantage of funding opportunities or collaboration that will help them develop and take products to market.